RESCUED BY C#

```
///  <summary>
///    Summary description of Welcome.
///  </summary>
public class Welcome
{

    public static void Main (string[] args)
    {

        MessageBox.Show ("Welcome to C Sharp!",
                         "Howdy");

    }

}
```

CHARLES WRIGHT

Kris Jamsa, PhD, MBA,
Technical Editor

ONWORD PRESS
THOMSON LEARNING

Australia • Canada • Mexico • Singapore • Spain • United Kingdom • United States

ONWORD PRESS

THOMSON LEARNING

Rescued by C#

Charles Wright

Kris Jamsa, PhD, MBA, Technical Editor

Business Unit Director:

Alar Elken

Executive Editor:

Sandy Clark

Senior Acquisitions Editor:

Gregory L. Clayton

Executive Marketing Manager:

Maura Theriault

Channel Manager:

Mary Johnson

Marketing Coordinator:

Karen Smith

Executive Production Manager:

Mary Ellen Black

Production Manager:

Larry Main

Art/Design Coordinator:

David Arsenault

Editorial Assistant:

Jennifer Luck

Full Production Services:

Liz Kingslien
Lizart Digital Design, Tucson, AZ

NOTICE TO THE READER

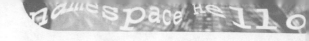

PREFACE

C# (pronounced "C Sharp") is a new object-oriented programming language introduced by Microsoft to support its new .NET frameworks. It is a component-oriented language that treats everything as an object. C# is a very easy language to learn for newcomers and bears a striking resemblance to Visual Basic and Java. If you are an experienced programmer and have used a language such as C++, you will find the transition to C# painless because of the similarity of syntax and keywords.

The concept behind the .NET framework is similar to the Java Virtual Machine. A C# program is compiled into an intermediate code, which then is translated by the .NET runtime to native code for the computer on which the program is running. The result is faster execution time and easier interfacing to programs written in other languages.

This book is intended for the beginning programmer. It is not a comprehensive course on C#, but it will teach you the basics and get you started programming in this new language. The book does not assume that you already know C++ or Visual Basic, but at times similarities, and some differences, with these other languages will be noted.

SOURCE CODE AVAILABILITY

Please visit www.onwordpress.com to download the source code for Rescued by C#. Instructions for downloading the source code will be available on the home page.

LESSON 1

```
/// <summary>
/// Summary description for Welcome.
/// </summary>
public class Welcome
{
    static void Main(string[] args)

        MessageBox.Show ("Welcome to c Sharp!",
                         "Howdy");

}
}
```

GETTING STARTED WITH C#

Whether you are a newcomer to programming or have some measure of experience, you are about to enter a new world of programming. You may know about object-oriented programming already, but C# (pronounced "C Sharp") is a language that treats everything—even number constants—as objects. From the time Microsoft introduced C#, there have been many attempts to compare it to C++. However, C# is not a new version of C++, so dispel any such preconceived ideas now. C# is related to C++ in about the same manner as Visual Basic is related to BASIC. Some of the keywords and syntax may be the same, but the languages are distinctly different. C# will not replace C++, nor will it replace Visual Basic. Instead, it may become another tool in your programming repertoire. This book will avoid any such comparisons, and will not assume that you have a working knowledge of C++.

C# is part of Microsoft's .NET programming environment. The .NET environment is *component-oriented,* and C# is intended to make writing components easier. Think of a component as a program that lets other programs access its functions. The .NET environment uses the Common Language Runtime interface when you run a C# application, and thus C# applications likely will be slower than applications that you write in C++. This lesson will introduce you to C# and the Visual Studio environment, in which you may write and test your programs. By the time you finish this first lesson, you will understand the following key concepts:

- Programming is the process of writing instructions for a computer using a language you can read and write, such as C#.
- Compiling is the process of converting the program you write into the ones and zeros the computer understands.
- You write and store your program in a source-code file. The C# compiler, in turn, will read the source-code file and convert the file's contents into instructions the computer understands, it then stores these instructions in another file called an *assembly*.
- When you run your C# program, the program launches the Common Language Runtime, which "manages" your code.
- C# is part of the Microsoft Visual Studio, which contains many tools to help you write and develop programs.
- Debugging is the process of finding and correcting errors in a program. To help you debug your programs, the Visual Studio contains a powerful debugger program.

Understanding Computer Programs and Programming

A computer program, or *software,* is a series of numbers your computer uses as *instructions* to perform a specific task. In the computer, these numbers are composed of 0's and 1's that represent the absence or presence of electrical signals. Different numbers activate different combinations of transistors, or gates, in the computer circuitry and cause the computer to perform certain operations. You create a program first by writing a list of *statements* in a programming language such as BASIC, C++, Pascal or C#. Then, another program called a *compiler* translates your statements into the numbers that are meaningful to a computer.

Learning to program is a matter of persistence and reinforcement. Few things worth doing are easy to master. I often compare the task of learning a new language to the process of trying to learn the Morse Code. It is more important that you study 15 minutes *every* day than that you study four hours a week, perhaps all in one day. Each day you review the Morse Code letters you already have learned, than tackle a few new ones. Similarly, in learning a programming language, you use the concepts you already have learned and tackle one or two new concepts.

As your skills advance, you will encounter "plateaus," periods when you just do not seem to advance as rapidly as you would like. You will find some programming constructs difficult to master. This is where persistence pays off. If you keep at it, eventually you will break out of a plateau and your skills will seem to rocket for a while. Keep tinkering and experimenting with a difficult construct and you will suddenly get an insight and exclaim, "Of course!"

C# is one of the tools used by a program called *Visual Studio,* which is Microsoft's offering of an *integrated development environment* (IDE). The package contains a source code editor, a compiler and a debugger, as well as other programming tools that you can run from the Visual Studio. You can perform all the tasks within the Visual Studio program, or you can write your programs and compile and run them from a command line prompt. (Compiling is the process of converting your source code into the binary instructions your computer understands). In the course of this book, you will do both.

Visual Studio.NET provides you with an environment in which you can write programs and components for Windows. To do that, of course, you need a basic understanding of programming. In addition, C# is a new programming language, and the .NET environment still is evolving. In this book, you will learn the basics of the C# language within the Visual Studio environment with the goal of placing you into a position to better understand how to write Windows programs.

Installing Visual Studio.NET

Before you begin, you should be aware that Visual Studio.NET requires an advanced computer to perform adequately. I have installed it on a 200-MHz Pentium class machine with 256 megabytes of memory running Windows 98, but its performance was far less than satisfactory. Microsoft recommends at least a 400-MHz Pentium II class machine with 256 megabytes of memory. My test computer is a dual 400-MHz Pentium II with 512 megabytes of memory running Windows 2000, and the performance is not stellar. In addition, the Visual Studio main window is a very busy and crowded place, and your monitor and video adapter should be able to handle a resolution of 1024 by 768 pixels. You can use a lower resolution, such as 800 by 600 pixels, but you probably will want to set most of the tool windows to "auto hide," as described later in this lesson.

When you insert the first CD, the setup program should start automatically. If it does not, you may run SETUP.EXE directly from the CD. Unless you are re-installing Visual Studio.NET, the setup program probably will inform you that you need to upgrade some Windows components. These include any service packs for your version of Windows, Front Page server extensions and the .NET software development kit (SDK). The setup program will ask you to insert the Windows Component Update disk into your CD.

While the setup program is installing the update, your computer may reboot several times. This is normal, and you should expect this stage to take 20 minutes or more, depending upon the speed of your computer and the number of components that the setup program needs to upgrade.

After the Windows Component Update is finished, re-insert CD-1 to install the Visual Studio.NET program. Shortly the setup program will prompt you to insert additional CDs.

When finished, you also will find setup has installed the MSDN library. This will be your help file for Visual Studio.NET, but this library is incomplete and still evolving. Depending on your version, you may encounter documentation errors, and a number of examples of sample code that may not compile or run. Considering Microsoft's record of keeping the MSDN up to date for previous editions of Visual Studio, you should use this library with some degree of skepticism. This does not mean that it is not a good source of information, but do not assume that because something is in the MSDN library it is correct.

Running Visual Studio

Visual Studio is a Windows program. When you install it, the installation program creates a group of items on the start menu that lets you run it from the desktop. Select the Start menu Program option. Windows, in turn, will display the Programs menu, within which you should see an entry for "Microsoft Visual.NET 7.0." Select this item and you get another menu list containing the Visual Studio.NET and the MSDN help program. Select the item labeled "Microsoft Visual.NET 7.0" to start the Visual Studio program. The Visual Studio window will appear, as shown in Figure 1.1.

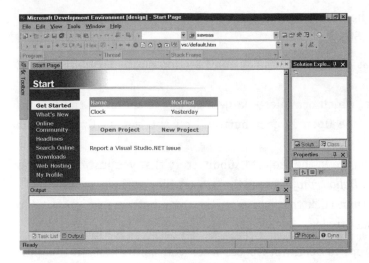

Figure 1.1. An empty Visual Studio as it first appears on your screen.

The Visual Studio is a busy place. Throughout this book, you will learn how to use the various windows and tools in the Visual Studio, but first you want to get to the point where you can write and launch your own programs. Think of the Visual Studio as the computerized equivalent of a workshop. You place various tools around the workshop and you might set aside certain areas for a specific task. In a workshop, for example, you might set up a table saw in the corner and that is where you do most of your work with wood. Another area might contain a lathe and drill press where you do your metal work. Similarly, each section of the Visual Studio has a purpose, and you will perform specific tasks in these areas.

Visual C# does not limit you to developing your programs in the Visual Studio, however. There will be times when you will want to write a utility program that you run from the command line. Lesson 4 will teach you how to create and edit program files, then compile and run them without ever entering the Visual Studio.

Loading a C# Program in Visual Studio

Visual Studio is designed to help you develop many different types of projects, not just C# projects. Within Visual Studio, you cannot just create a C# program file and expect to compile and run it. Instead, you first must create a project. The project defines the tools and file types your program will use, and determines how the Visual Studio handles your source files.

You will build a project in the next lesson, "Building, Running and Saving your First C# Program," and you also will learn about the files the Visual Studio creates for you and how it uses them. For now, you can download the sample program from this book's Web site as discussed in this book's introduction.

Run the program. When the program ends, use the following steps to open the project in the Visual Studio:

❶ Start Visual Studio as described in this lesson's section entitled *Running Visual Studio.*

❷ From the Start Page, which occupies a large area in the center of the screen, click the mouse on the Open Project button. You will get an Open Project dialog box.

❸ In this dialog box, change to the Project1 subdirectory that you just created and look for a file named *Hello.csproj*. This is the project file for the program.

❹ Select this file and then click the mouse on the Open button. The project file will load in the Visual Studio.

You will use these steps to open workspaces when you want to return to a program to work on it. The Visual Studio will create new workspaces in separate directories. To open a different program, simply navigate to the directory and open the workspace file. The Visual Studio will open any projects that you have created in the workspace at the same time.

Within Visual Studio, locate the Build toolbar as shown in Figure 1.2. If the toolbar is not visible, right-click your mouse in a blank area to the right of any toolbar. Visual Studio, in turn, will display a list of toolbars. Toolbars that are visible will have a check mark next their names. Within the toolbar, select Build from the list to make it visible.

Build
Selection Build Cancel

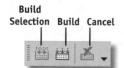

Figure 1.2. The Build toolbar contains the buttons you need to build your program

Building your program involves running a program—the compiler. As briefly discussed, the program reads your source files and converts them to the numbers that correspond to instructions the computer understands. The compiler then writes this converted code to another file with an extension of *.exe*. This file is your *executable* program.

If you have not opened the Lesson01 project as previously described, do so now. Next, to build the program, click your mouse on the Build toolbar Build Program button. You also may select the Build menu, then the Build item on that menu, or simply type **Ctrl+Shift+B.** Visual Studio will begin building the project and in a few seconds a line similar to the following will appear in the Output window at the bottom of the screen:

```
Build: 1 succeeded, 0 failed, 0 skipped
```

When this message appears, the program is ready to run or debug. *Debugging* is the process programmers perform to test their program for errors and then to fix the errors. The Visual C++ debugger lets you run your program one line at a time to test your code.

Next, locate the Debug toolbar, which is shown in Figure 1.3. If the Debug toolbar is not visible, use the process of right-clicking your mouse on a blank area of the frame to display a list of toolbars, then select Debug from the list.

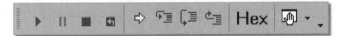

Figure 1.3. The arrow at the far left side of the Debug toolbar is the Start button. Use this button to start your program in the Visual Studio debugger.

Within the Debug toolbar, click your mouse on the Start button to start the program in the debugger. Shortly, your program displays a message box with the words "Welcome to C Sharp!" as shown in Figure 1.4. Within the message box is a button with the word "OK" on it. Click your mouse on this button to close the message box and exit the program.

Figure 1.4. When you build and run the hello.exe program, your screen will display a window containing a welcome message, using a Windows message box.

Although the message box may appear simple, the program that displays the box illustrates several steps you will perform frequently within the Visual Studio to generate the project. The listing that follows shows the complete code for the program. In the next lesson, you will go through the steps of actually creating this program:

```
namespace Hello
{
    using System;
    using System.Windows.Forms;
    /// <summary>
    ///     Summary description for Welcome.
    /// </summary>
    public class Welcome
    {
        public static void Main(string[] args)
        {
            MessageBox.Show ("Welcome to C Sharp!",
                             "Howdy");
        }
    }
}
```

Printing a Program File in C#

Visual Studio allows you to print a file from within the IDE, or to print just a portion of it. You will find the partial-print option handy when you only need to print a specific section of your source code. To print a file from the Visual Studio, perform these steps:

❶ Select the file in the client area by clicking your mouse on the tab that contains the file name. Visual Studio will respond by bringing the file to the front of the display.

❷ Select the File menu Print option (if you have not selected the client area as discussed in Step 1, the Print item will be "grayed out" and you cannot select it). Visual Studio will display a Print dialog box with the name of your default printer highlighted. Notice there is a button labeled "Selection" in the Print Range area, but it is disabled.

❸ Within the Print dialog box, click your mouse on the OK button. The Visual Studio will begin sending the file to the printer and the Print dialog box will disappear.

To print just a portion of a file, select the file as in Step 1 above, and then perform these steps:

1 Place the caret (the window cursor) in front of the first character of the portion you want to print.

2 Hold the Shift key down. While holding the Shift key down, use the arrow keys to move to the *last* character in the portion you want to print. Visual Studio will highlight the text you select using reverse video.

3 Select the File menu Print option. Do not click on the editing window again or move the caret. If you do, you will cancel the selection. Visual Studio will display the Print dialog box.

4 Within the Print dialog box, notice that the Selection item is enabled and Visual Studio has selected it for you. To print just the text you highlighted, click your mouse on the OK button. If you change your mind and decide to print the entire file, click the left mouse button on the All button before clicking OK.

Setting Up Visual Studio

If you do not like the way the Visual Studio appears or the location of the various windows and toolbars, you can move them around to suit your taste. Further, you can move all the toolbars to other places around the studio by grabbing the left side of them with the left mouse button and dragging them to where you want them. You can leave the toolbars "floating," unattached to the window frame as tool windows.

Figure 1.5 shows two views of the Visual Studio. The left view has all the toolbars attached to the Visual Studio frame, and the right view shows them floating. With the exception of the Menu Bar, you may hide all the toolbars when you do not need them. You can set other tool windows—such as the Output window, the Solution Explorer and the ClassView Window—to float or to "auto hide." The right view of Figure 1.5 shows the Solution Explorer floating. When you set them to auto hide, they shrink into a tab at the side of the window frame and reappear when you move the mouse over the tab.

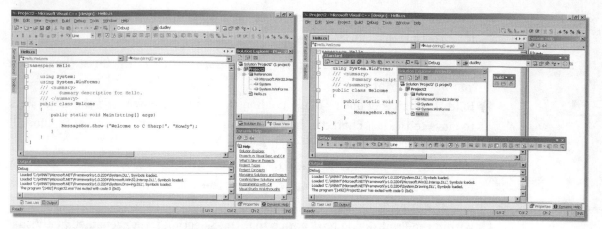

Figure 1.5. You can detach any of the toolbars or tool windows in the Visual Studio and move them where you want them. The left view shows all the toolbars attached. At right, the Standard, Debug and Build toolbars and the Solution Explorer window are floating.

To hide or show a toolbar, right-click your mouse on a blank area of the frame (the area to the right of the toolbars is a good place). The Visual Studio, in turn, will display a selection menu. Within the menu, each visible toolbar will have a check mark next to its name. Using your mouse, click the button to turn a toolbar's display on or off.

You should practice using this menu and make yourself familiar with the various toolbars. You will need this knowledge when you start debugging your programs.

You can also move or hide the specialized windows, such as the Output window and the Solution Explorer. To select options for a tool window, right-click your mouse on the title bar. Visual Studio, in turn, will display a menu you can use to select the settings you desire.

Saving Your Program Code

To reduce your risk of losing work, you may want to configure the Visual Studio to save your program files automatically when you compile a file or build your program. As you develop more complicated programs, this is important because Visual Studio is not a perfect program and it has been known to crash during debugging sessions. A crash is caused by an unexpected error in a program that makes it impossible to continue.

Select the Tools menu Options item. Visual Studio, in turn, will display the Options dialog box, which contains many pages you can use to set options throughout the Visual Studio. To display a page within the dialog box, select an item from the tree on the left side of the dialog box. Right now, for example, you are interested in the Environment options. Select that page by clicking your mouse on it. From the list that appears below the Environment item, click your mouse on the Projects and Solutions item. The dialog box display should be as shown in Figure 1.6.

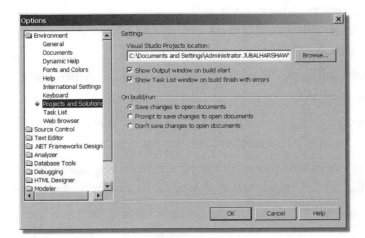

Figure 1.6. Displaying pages within the Options dialog box.

Within the middle of the page is a group called "On build/run." Select the first item, "Save changes to open documents." Make sure this box is checked and the Visual Studio will save your files automatically when you compile or build your program.

Saving Your Files

To save a file without running the compiler, make sure you have selected the window containing the file in the client area, then select the Save item on the File menu. The Save item will include the name of the file you have selected. For example, if you selected the "Hello.cs" source file, this item will read "Save Hello.cs." You also can select Save All to save all the files that you have modified since your last save operation.

When you exit the Visual Studio or close a workspace, the Visual Studio will check whether you have modified any files and not yet saved them. If it finds any, it will display a list of the files that have changed since the last save operation and ask whether you want the changes saved.

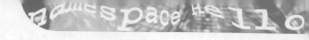

WHAT YOU MUST KNOW

In this lesson, you learned how to install and run the C# compiler and how to set up the Visual Studio to save your program files when you build your program. You also learned how to view and print your program source code from within the Visual Studio. In Lesson 2, "Building, Running and Saving your First C# Program," you will create your own programs using Visual Studio. Before you continue with Lesson 2, however, make sure you have learned the following key concepts:

- When you write a program, you specify instructions for the computer in a language you can understand. Visual Studio is a tool that lets you write programs in C# and other languages.

- After you have written your program in C#, a program called a compiler translates it from C# to the ones and zeros that the computer understands.

- You keep your program code in a source file. The C# compiler reads this source file and converts it to the number instructions the computer can understand.

- You find and correct errors in your program through a process called debugging. The Visual Studio contains a debugging tool that helps you through this process.

- Like most Windows programs, the Visual Studio contains commands that let you print your source files.

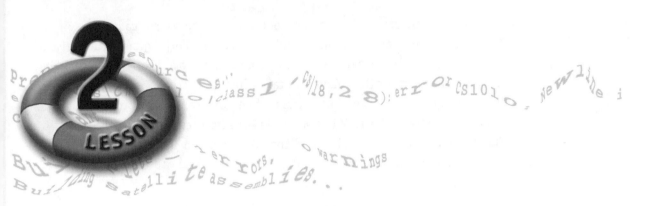

BUILDING, RUNNING AND SAVING YOUR FIRST C# PROGRAM

V isual Studio.NET is an *integrated development* environment (IDE). It is a program that you use to write, compile and test your programs. However, you cannot just create a program file and expect the Visual Studio to know what you intend to do with that file. To give the Visual Studio that information, you must create a *project*. During the process of creating a project, you give the Visual Studio information about the type of project you want to create—such as a Windows project, an application for the World Wide Web, or a Windows control program—and the language in which you want to write that program, such as Visual Basic or C#.

From the information you supply, the Visual Studio creates support files containing information about your project. The Visual Studio uses these files when you open and work with your project. In this lesson, you will begin creating projects in the Visual Studio, starting with a blank project, which you will transform into a Windows program by changing options in the Visual Studio. By the time you finish this lesson, you will understand the following key concepts:

- Before you start writing a program using the Visual Studio, you must create a project. The project file contains information the Visual Studio needs to compile and debug your program.
- Visual Studio supports several other languages in addition to Visual C#. From the C# options, you may select several different project types.

- The Visual Studio stores the program code you write in source-code files.
- To test run a program, you must build, or compile, the program within Visual Studio. Any time you make a change to your source-code files, you must build the program again for the changes to take effect.
- If the compiler encounters any errors while compiling your program, it will report the errors in the Task List window at the bottom of the Visual Studio.
- The Visual Studio debugger is a tool that will help you to find and correct programming errors in your code.

Creating a Project

To create and write a program using the Visual Studio, you first must create a project. The information you supply during the process of creating a project gives the Visual Studio the information it needs to help you work with your program. This information informs the Visual Studio what tools to use as you build your project and what rules to apply to the files you create, such as the *syntax* or grammar rules for the programming language you want to use.

In this section you will create an empty project, which you will use to create a *source-code* file to contain the programming statements to write a message to your screen. At first, the project will simply write a message to the *console*—the MS-DOS prompt—and you later will change project options until it produces the message box you encountered in Lesson 1, "Getting Started with C#." In later lessons you will learn how to create other types of projects, including Windows programs based on *forms,* the windows and dialog boxes that make up an application's *user interface.*

Start Visual Studio as described in the section on Running Visual Studio in Lesson 1, "Getting Started with C#." To create a project, select the File menu, then the New item and finally the Project item (or use the menu short-cut by typing **Ctrl+Shift+N**). If the Start Page is visible (such as when you first run Visual Studio), you may click your mouse on the New Project button of the Start Page. Visual Studio will display the New Project dialog box, as shown in Figure 2.1.

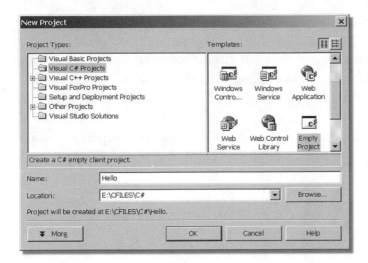

Figure 2.1. The New Project dialog box lists the languages and project types that are available along with a set of icons that represent projects that you may create using Visual Studio.

In the Project Types section of the New Projects dialog box, select Visual C# Projects. The Visual Studio will display a set of templates representing C# project types. For now, select Empty Project in the right side of the dialog box.

Within the Name field, type **Hello.** Next, within the Location field, enter the directory in which you want to create and store the project. Visual Studio will respond by showing you the project directory just below the Location field.

Click your mouse on the OK button to create the new project. You may not notice a lot of changes to your screen. If you examine the Solution Explorer window on the right side of your screen, you will see that Visual Studio has added a line containing your project name, as shown in Figure 2.2.

Figure 2.2 The Solution Explorer should list any projects you have open in the Visual Studio.

The Solution Explorer connects you to the various parts of your program such as files containing code, forms that your program will display, or modules that your program will reference. Right now your project is empty, so you must add some code to make it perform some action.

In C#, all code is contained in *classes*. A class is the primary mechanism for object-oriented programming in C# and other languages such as C++. Think of a class as a container—a "parts bin" if you want to continue the workshop analogy—that holds related *functions* and *variables* in an object-oriented program. You will learn more about classes and their components in later lessons.

To add a class to your program, right-click your mouse on the name of your project in the Solution Explorer window (the line that displays "Hello") to display a menu of options. Select the Add item and Visual Studio will display a submenu. At the bottom of the submenu is an item labeled "Add Class." Select the Add Class item to display the Add New Item dialog box as shown in Figure 2.3.

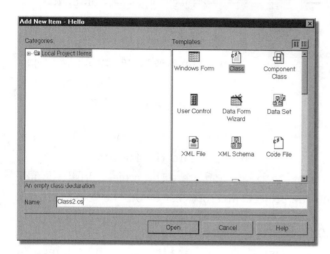

Figure 2.3 You may insert new objects into your program using the Add New Item dialog box.

Within the Categories pane of the Add New Item dialog box, select Local Project Items. Within the Templates pane, select the C# Class item.

The Name field at the bottom of the Add New Item dialog box contains the name of the file that will hold your new class. In current versions, the file name will be the same as the name of your new class, followed by the *.cs* extension to indicate a C# program file. Microsoft may change this in future versions of C#, but for now the class name and the file name must be the same.

You can use the default names for the class (Class1) and file (*Class1.cs*). Click on the Open button and the Visual Studio will create the *Class1.cs* file and display it on your screen.

Your program is not ready to compile and run yet, so now would be a good time to examine the Visual Studio and the files it has created for your project.

Understanding Your Program Files

To save information about your project and the program it contains, the Visual Studio creates several files in your project directory. If you start Windows Explorer and navigate to the project directory (the location you entered in the New Project dialog box in the previous section), you will see a listing of these files. Until you have more experience, you should not edit these files directly.

Instead, let the Visual Studio maintain the files and make any necessary changes.

First, you will see a file named *Hello.csproj*. This is the C# project file. When you close your project, Visual Studio will save the settings in the *csproj* file. Later, when you return to work on your program, Visual Studio will apply the settings when you re-open the project.

The *Hello.sln* is the "solution file." If you have used previous versions of Visual Studio, the solution file is similar to the "workspace" file. The solution file contains information about your project, and a single solution file may contain information about several projects.

The *Hello.suo* file contains information about the contents of the various tool windows such as the Class View panel. It also contains the Visual Studio options you have selected using the Options item on the Tool menu.

Starting the Compiler

Before you can compile and run your program, you must give the program an *entry* point—the position at which it will begin executing your program code. Every Visual C# program must have an entry point called *Main()*. When your program starts, Windows transfers control to the first statement in the *Main()* function. (A statement is a single instruction in your program and a function is a block of related statements. The open and close parentheses following a name indicate that it is a function.) Add the *Main()* function to the code that Visual Studio created as shown in the following listing:

```
namespace Hello
{
    using System;

    /// <summary>
    ///     Summary description for Class1.
    /// </summary>
    public class Class1
    {
        public Class1()
```

```
    {
        //
        // TODO: Add Constructor Logic here
        //
    }
    static void Main()
      {
        Console.Write ("Hello C Sharp!\n");
      }
  }
}
```

There are three ways to compile and build your program. First, compile your program by clicking on the Build toolbar Build button (Lesson 1 presented the Build toolbar). Also, you may type the menu short-cut **Ctrl+Shift+B** to compile the program. Third, select the Build menu Build option.

The Visual Studio will start the C# compiler and build your program. After a few seconds, a message similar to the following should appear in the Output window at the bottom of your screen:

```
Build: 1 succeeded, 0 failed, 0 skipped
```

Using the Output Window

Between the time you start the compiler and the time the build message appears, the compiler reads your source file and converts it into the number codes that the computer can understand. The compiler then writes the numbers to another file—the executable, or program, file.

If the compiler encounters any errors in your program code, it will write the errors to the Output window at the bottom of the Visual Studio screen. (The bottom of the screen just below the source-code files contains a window that may contain any of several windows. One of these windows is the Output window. If the Output window is not visible, look for a tab labeled "Output" and click on this tab to display the Output window. Later in this lesson, you will learn about other windows that may appear in this area.)

To demonstrate error output, introduce an error into your program code. Remove the closing quotation mark from the call to the *Console.Write()* function so that it reads as shown here:

```
Console.Write ("Hello C Sharp!\n);
```

Now compile your program as described in the last section. Instead of the successful build message, you will see the following lines. You may have to scroll through this window to see all of the text in the Output window.

```
— — — Build started: Project: Hello, Configuration: Debug .NET

Preparing resources...
e:\cfiles\c#\hello\class1.cs(18,28): error CS1010: Newline
in constant

Build complete — 1 errors, 0 warnings
Building satellite assemblies...

— — — — — — — — — — Done — — — — — — — — — —

    Build: 0 succeeded, 1 failed, 0 skipped
```

The compiler could not complete the process of converting your source code into an executable file, so it has issued a notice that the build failed. This is an example of a *syntax error,* which means that a statement was not properly entered according to the punctuation rules of the C# programming language. Such errors are like typographical or spelling errors when you write a letter in a word processing program.

Before you correct the error and continue, take a look at the results a little more closely, as discussed in the following section.

Examining the Results

Within the Output window, locate the line that tells you there is a "newline in constant." When you double-click the mouse on the line, the Visual Studio will locate the error in your source code and place the caret at the beginning of the text containing the error. (See the hint "Caret vs. Cursor" for an explanation of the difference between the caret and the mouse cursor).

Sometimes the Output window will contain more than one error. You can locate each error by double-clicking your mouse on the error line and the Visual Studio will move the caret to the line containing the error.

In addition, Visual Studio will prepare a "task list" for you listing the errors. To switch to the task list, click your mouse on the tab marked "Task List" at the bottom of the Visual Studio screen. (The Task List is another window that might appear in the same location as the Output window.) Figure 2.4 shows the window containing the Output and Task List tabs.

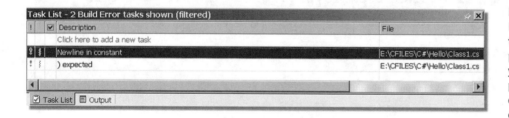

Figure 2.4. The Task List window lists the errors the compiler finds when you compile your project. Double-click your mouse on a task to display the source file and move to the error.

Notice that the Task List contains two errors, but the Output window showed only one error. A single mistake in a program statement may cause the compiler to generate more than one error because it temporarily loses track of the *syntax* of your program. When you get more than one error, always work on the first error in the list. Correcting the first error may also correct one or more of the following errors.

To move to the line containing an error, double-click your mouse on the error within the Task List and Visual Studio will move to the line in your source-code file containing the mistake.

After you correct the error, you must recompile your program for the changes to take effect. If the Visual Studio displays no errors when you recompile your program, you are ready to test-run the program in the debugger.

Caret vs. Cursor

In the pre-Windows days, programs such as text editor or even the command screen might have a single indicator—such as an underscore, a vertical bar or even a solid block—that indicated where the next character you type would appear. Programmers and users commonly referred to this mark as the "cursor."

With the introduction of Windows, however, that situation got a little more complicated. There was a mark where the next character typed would appear, and there was another mark to indicate the current mouse position.

In Windows programming, the mark that indicates the mouse position is the "cursor" and might be a pointer arrow, an hourglass or any of several other common symbols.

The position in a text window where the next character you type will appear is the "caret." The caret might be a vertical bar, and underscore or a block that is superimposed over a character.

Running Your Program

Not all errors are syntax errors. Sometimes you will encounter programming errors that will compile but will cause your program to run incorrectly. Debugging is the process of test-running your program in a protected environment such as the Visual Studio debugger so that you may locate and correct these errors.

To run the *Hello* program in the Visual Studio debugger, press the **F5** key, or click your mouse on the Debug toolbar **Start** button.

When you run your console program, notice that it runs and then exits immediately, giving you very little time to see the output. To stop the program before it exits so you can see the text, click your mouse on the curly brace "}"

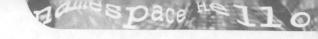

immediately after the line that calls the *Console.Output()* function, then press **Ctrl+F10** or select the Debug menu Run to Cursor item. The debugger, in turn, will run the program up to the curly brace and then pause. At this point, notice that a DOS prompt icon appears on the Windows Task Bar. Click on this icon to examine the output line:

```
Hello, C Sharp!
```

Using Breakpoints to Pause Your Program

To halt the program at a particular line in the debugger, you may set a breakpoint in the program code. A breakpoint tells the debugger that it should pause the program at a particular line or when a particular condition occurs.

To set a breakpoint, click your mouse in the margin bar at the left side of the source-code window at the line where you want to pause your program. The Visual Studio will place a red circle in the margin to indicate the breakpoint and highlight the text on the line, as shown in Figure 2.5.

```
/// </summary>
class Class1
{
    public Class1()
    {
        //
        // TODO: Add Constructor logic here
        //
    }
    static void Main(string[] args)
    {
        Console.Write ("Hello, C Sharp!\n");
    }
}
```

Figure 2.5. The Visual Studio places a red dot in the margin where you click the mouse to indicate a breakpoint.

To remove the breakpoint, simply click your mouse on the red dot. The Visual Studio will respond by removing the dot and the breakpoint.

Breakpoints are handy for debugging larger programs. As you will learn later, you can place conditions on the breakpoint, causing the program to pause only when a variable contains a certain value.

When you add breakpoints to your program files, you may display and edit them—and set conditions—by showing the Breakpoint Window. To do this, select the

Debug menu, then the Windows item. From the submenu that appears, select Breakpoints. The Breakpoint Window will appear on a tab on the window at the bottom of the Visual Studio screen. This is the same window that contains the Output and Task List tabs.

At this point, your project is a "console program," meaning that it first must open an MS-DOS console window before producing any output. Ultimately, of course, the purpose of Visual Studio is to help you to write programs for Windows.

You can change your console program from one that writes to a console screen to one that writes to a Windows message box simply by changing the output line in your program. To do this, you will perform two steps: first, you will convert it to a console program that opens a console window, then generates a message box window. Then, you will change the program to one that bypasses the console window altogether and displays only a message box.

For the first part, you will find the Solution Explorer window on the upper right side of your screen. If it is not visible, look for a labeled "Solution Explorer" and click on it. (The entire label might not be visible depending upon the resolution of your screen and the size of the window, so the label might read something like "Solution Ex..."). The Solution Explorer is shown in Figure 2.6.

Figure 2.6. The Solution Explorer window with the project and references list expanded.

To view the options, you must expand the list fully. If the line containing the project name "Hello" contains a "+" symbol next to it, click on the symbol to expand the project list. Repeat this step to expand the References portion of the list.

Notice that your program already contains several "references." Throughout the documentation, you will see the term *reference* used in several different ways. In this context, a reference is a code module that your program uses. By using references instead of adding the code module to your program code, the module may be located anywhere, even on a different computer in a network environment. To use the message box, you will need to add a reference to the System.Windows.Forms module.

Within the Solutions Explorer window, right-click your mouse on the References line. Visual Studio will display a pop-up menu with a single item, Add Reference. Select this item to display the Add Reference dialog box.

Search through the list on the Add Reference dialog box for an item labeled "System.Windows.Forms.dll" in the Component Name column. Select the item, then press the **Select** button on the right side of the dialog box. The item will appear in the Selected Components list at the bottom of the dialog box. Click on the OK button to add the reference to your project. After the dialog box closes, you should see the System.Windows.Forms object in the reference list of the Solution Explorer.

Next, modify the Class1.cs source-code file as follows. Notice that you are adding the System.Windows.Forms *namespace* to the code as well as changing the output in *Main()* from *printf()* to *MessageBox.Show()*. (You will learn more about namespaces in Lesson 4, "Understanding a Few Key C# Basics.")

```csharp
namespace Hello
{
    using System;
    using System.Windows.Forms;
    ///     <summary>
    ///         Summary description for Class1.
    ///     </summary>
    public class Class1
    {
        public Class1()
        {
            //
            // TODO: Add Constructor Logic here
            //
        }
        static void Main()
        {
            MessageBox.Show ("Hello C Sharp!\n", "Howdy");
        }
    }
}
```

Build and run your program once again. This time, the program will open a console window, then display a message box.

Finally, to remove the console window from your program, you must change the application type from "Console Application" to "Windows Application." To change the project type, right-click your mouse on the project name in the Solution Explorer (the line that reads simply "Hello"). From the menu that pops up, select Properties.

When the Hello Property Pages dialog box appears, select Common Properties, then General from the list on the left side of the dialog box. From the list in the large window on the right, click on Output Type and a down arrow will appear to the left of the words "Console Application." Click on the down arrow to display a list of output types. From this list, select "Windows Application." Finally, click the OK button to close the dialog box.

Recompile and run your program once again. This time the program will skip the console window altogether and display only the message box with the phrase "Hello C Sharp!"

The code for the basic console program can be found on this book's Web site as discussed in this book's introduction. The modified version that displays only the message box is in the Lesson01\HelloWin directory.

Note: *You should remember that because a program loads and runs on your computer, this does not mean that you can copy the executable to another computer and expect it to run. You have seen that you can access other code modules using references to those modules. If they are not installed on another computer, your program will not be able to run. Installing your program and the code modules that it needs to run on another computer are part of the* deployment *process in the Visual Studio.*

Closing and Saving Your Program

Unless your project is small and relatively simple, you likely will not be able to complete all your programming in a single session with the Visual Studio. Instead, you will close your project from time to time, then return to it later to continue working on your project.

At other times, you may have completed work on a single source-code file and you want to remove it from the screen. Visual Studio is a very busy place, and removing clutter to keep your workspace clean is important.

To close a single source-code file, select it in the editing area, then select the File menu Close option. If you have made changes to the file, Visual Studio will ask whether you want to save those changes before closing the file. Notice that the File menu also includes an option to save the contents of the selected file. You may select the Save item at any time to save the source-code file without closing it, or you can simply type **Ctrl+S.**

When you exit Visual Studio by selecting the File menu Exit option, Visual Studio will ask you whether you want to save changes to your files before exiting.

To close a project completely without exiting the Visual Studio, select the File menu, then select the Close Solution item. Again, if you have made changes to any of the files, Visual Studio will ask whether you want to save the files before closing the solution.

WHAT YOU MUST KNOW

In this lesson, you have learned a number of concepts about programming in the Visual Studio environment. You have learned how to create new projects, how to compile, and how to run programs in the Visual Studio debugger. You also have learned how to add references to your project to access code modules. In Lesson 3, "Exploring the Visual Studio," you will continue looking at different features and tools in the Visual Studio environment. Before you continue with Lesson 3, however, make sure you have learned the following key concepts:

- **x** When you write a program, you are writing instructions for the computer in a language you can understand. Visual Studio is a tool that lets you write programs in C# and other languages.

- **x** After you have written your program in C#, a program called a compiler translates it from C# to the ones and zeros that the computer understands.

- **x** You keep your program code in a source file. The C# compiler reads this source file and converts it to the number instructions the computer can understand.

- **x** You find and correct errors in your program through a process called debugging. The Visual Studio contains a debugging tool that helps you through this process.

- **x** You can access other code modules by adding "references" to them in your program. References allow you to place the code modules in any directory on your computer, or on any other computer on a network.

LESSON 3

EXPLORING VISUAL STUDIO 7

In the previous two lessons, you have learned how to access and use some of the built-in features of the Visual Studio.NET development environment. You have learned how to create a basic project and how to make some limited modifications to your program. Soon you will begin developing Windows programs, but first you need to learn more about the development environment. Taking a few minutes to learn how to use your tools will speed up your program development efforts. In this lesson, you will learn how to set options in the Visual Studio, how to use some of the tool windows and how to access the Visual Studio help system. You also will learn how to use the Visual Studio toolbars. By the time you finish this lesson, you will understand the following key concepts:

- Visual Studio options let you determine how the development environment will appear when you first start the program.
- The Visual Studio project options enable you to create several different project types using C#.
- Toolbars are buttons that represent many of the menu items and speed up your programming tasks. You may hide, show and arrange the toolbars to suit your needs.
- The Microsoft Developers Network is the help system for the Visual Studio. You may access help from many of the Visual Studio windows.
- The Find in Files command searches all the source files in your project for text that you specify.

Setting Visual Studio Options

Although you have worked only with simple projects, you have used a number of features of the Visual Studio tool windows. You have seen that Visual Studio contains a number of tool windows, so many in fact that many of them have to share the same real estate on your screen. To share screen space, Microsoft has placed many of the tool windows on tabs so that you can hide one window and reveal another simply by clicking on a tab.

The Visual Studio tool set includes an extensive dialog box to configure the layout and operation of many of the tool windows. To access the Options dialog box, select the Tools menu Options item. Visual Studio will display a dialog box similar to that shown in Figure 3.1.

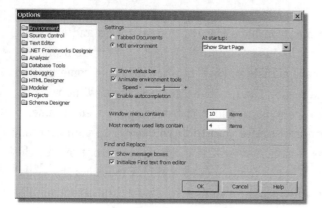

Figure 3.1. The opening view of the Visual Studio Options dialog box contains fields for setting general environment options.

On the left side of this dialog box is a tree containing entries for the various components of the Visual Studio environment. If you click your mouse on a folder in the tree, Visual Studio will expand the list to show additional configuration items. Clicking your mouse on the same item a second time directs Visual Studio to collapse the list to show only the folder-level items.

Some items contain additional folders that contain even more options. For example, expanding the Text Editor entry reveals a set of folders to set options for all the languages available in Visual Studio, or to set options for each individual language.

The first page, the Environment page, contains general options for the Visual Studio. Selecting between Tabbed Documents and MDI Environment, for example, determines how your source-code windows will display. If you select Tabbed Documents, then individual source-code files will be arranged so that you select them by clicking on a tab in the editing area. Selecting MDI Environment arranges the windows in the standard *multiple document interface* style, in which you may cascade the windows or tile them using options on the Window menu. With either Tabbed Documents or MDI Environment you may step through documents using the Control and Tab keys. Clicking the Reset Window Layout button will return all the settings to their default values.

Unchecking the Show Status Bar box will hide the status bar at the very bottom of the Visual Studio. The status bar contains panels that show you the column and line number where the caret is located. This is useful during debugging, but if you do not need this information you might consider hiding the status bar. Considering density of tool windows and toolbars, hiding the status bar would give you a little more space for displaying your source-code files.

Animate Environment Tools determines whether the tool windows you set to auto hide appear to slide out from the edges or simply pop up when you select them. If you select this box, the speed bar just below the box determines how quickly the tool windows will slide. The speed range is from an irritatingly slow to a just barely tolerable fast. To me, the automation adds nothing to a development environment, so I leave the box unchecked, making the tool windows pop up and hide almost instantly. Actually, unchecking the box does not disable the animation; it just makes it so fast that the tool windows appear and disappear very quickly.

You should spend some time familiarizing yourself with the Options dialog box. You can get a detailed explanation of each page by selecting an item in the tree, then clicking on the Help button.

Using the New Project Dialog Box

The starting point for any project in Visual Studio.NET is the New Project dialog box. The Visual Studio is an *integrated* development environment, and from this dialog box you can select the language you want to use for your application. Visual Studio supports several languages, including Visual Basic and Visual C++. Of course, this book is concerned with C#, but much of what you learn about the various tool windows and toolbars will apply to other languages as well.

To display the New Project dialog box, select the File menu New option. Visual Studio, in turn, will display a submenu. From within the submenu, select Project. As a shortcut, you may type **Ctrl+Shift+N**.

The upper half of the New Project dialog box contains two windows. On the left is a tree containing the Project Type. This is where you will select the language you will use for your project.

On the right is a list of the templates that are available for the project type you select in the Project Type tree. The contents will change as you select other languages or project types. You can display this list in two different ways by selecting one of the buttons just above and to the right of the list. The first button displays the list in large icon mode. The second button displays the list in small icon mode. These modes are the same as those you see when you select a view from a directory in Windows. Figure 3.2 shows the C# templates in both large and small icon modes.

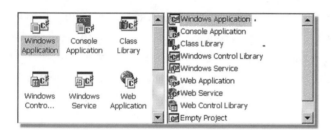

Figure 3.2. The left side shows the C# project templates in large icon mode; the right side shows the same templates in small icon mode.

Directly below the Project Type and Template windows is a static text field that gives a one-line description of the template you select in the right window. For example, selecting Windows Control Library in the template window displays "Create a C# Windows Control Library (DLL)" in this field. A one-line description is not very telling, but you can summon a fuller description by clicking your mouse on the Help button.

The next item on the dialog box is the Name field, where you will enter the name of your project. The Visual Studio will create a directory using the project name, and your project files will use the same name.

Below the Name field is a box in which you may enter the directory in which Visual Studio will create your project. The full path of the project will be the directory name you enter here plus the project name. According to the MSDN documentation, Visual Studio will create new projects in the location you select in the Options dialog box by selecting Environment, then Projects and Solution.

However, the default path Visual Studio selects will vary according to the project type and template you select.

Selecting a Project Type

Your first task in creating a project is to select a project type and template from the New Project dialog box. For now, make sure the Visual C# Projects item is selected. Later, you will need to use the Setup and Deployment Projects item to create a setup program for your applications, which will allow you to install the applications on other computers.

Table 3.1 summarizes the various templates available when you select C# project types.

Template	Description
Class Library	Create classes for use in other applications.
Console Application	Create a command-line application.
Empty Project	Create an empty solution file for a local application.
Empty Web Project	Create an empty solution file for a Web-based application.
ASP.NET Web Application	Create an application with Web pages as a user interface.
Web Control Library	Create a project for controls to use in Web applications.
Web Service	Create a project to build a service to use from other applications.
Windows Application	Create an application with a Windows form-based interface.
Windows Control Library	Create a project to hold controls for use in other Windows-based applications.
Windows Service	Create a project for a Windows-based service.

Table 3.1. Templates available for C# projects.

It is important that you understand the difference between a "solution" and a "project." A solution is a container for Visual Studio.NET projects. Conceptually, think of it as an office in which there are several desks. The office might be the accounting office of a firm, and one desk might be where a clerk works on accounts payable, another might be where a clerk works on accounts receivable, and yet another might be the payroll desk. The desks are the projects, and together they make up the solution, the accounting department.

A project is a collection of components—which may be source-code files, dialog boxes or images—that serve a single purpose. There may be one, two or even more projects in a single solution.

A solution also may contain components that are not a part of a project. In the hypothetical accounting office, there might be a number of file cabinets that hold files common to the functions of more than one desk. Thus, these components might be the elements that are common to all the projects.

Using Visual Studio Windows

Earlier versions of Visual Studio had neat names that you could attach to the tool windows spread around the window frame. For example, there was the Workspace Window, which contained several different views of a project or workspace.

Visual Studio.NET has lots of windows, but it doesn't have the tags that you can assign conveniently to the various places around the frame. Further complicating the issue is the fact that seemingly unrelated windows occupy the same area of the screen, adding themselves as tabs to other windows when you open the windows.

Actually, these windows provide for a high degree of flexibility in setting up Visual Studio. Open one of the projects that you created in an earlier lesson; the *Hello* project from Lesson 2, "Building, Running and Saving your First C# Program," is a good example.

If you have not already changed Visual Studio's default configuration, the project should open with the Solution Explorer and the Class View mounted on tabs in a window at the right middle of your screen. Below that, you will find another window with Properties and Dynamic Help on separate tabs.

Move your mouse cursor to the title bar of the Solution Explorer window (or the Class View window if that tab is selected). Right-click your mouse on the title bar and examine the menu that pops up as shown in Figure 3.3. This menu is common to most of the windows around the Visual Studio frame.

Figure 3.3. The menu that pops up when you right-click gives you options on how to display the window. A check mark next to an item means that item is selected.

Figure 3.3 shows the Dockable option selected. Click your mouse on this item to unselect it, and the Solution Explorer and Class View windows move to tabs in the center of your screen, along with the tabs for your source-code files. On the right side of the screen, the Properties and Dynamic Help windows have expanded to fill the empty space. Right-click your mouse on the Solution Explorer tab and reselect the Dockable item, and the window will move back to the right side of the screen. Do the same for the Class View tab and it will rejoin the Solution Explorer on the right.

Once again right-click on the Solution Explorer title bar and select Auto Hide. Move the mouse cursor far enough away and the Solution Explorer and Class View shrink into tabs on the far right side of the screen. Repeat the process with the Properties or Dynamic Help windows and they, too, become tabs, greatly expanding the area you may use for editing your source code, as shown in Figure 3.4. Yet any time you need one of the windows you need only click on a tab or let the mouse cursor hover over a tab for a second or so, and the window will pop back into view.

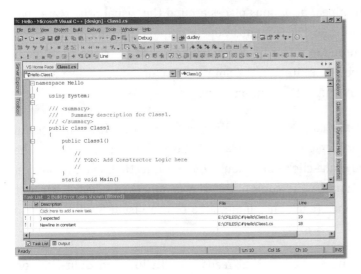

Figure 3.4. Setting the tool windows on the right side of the screen to Auto Hide greatly expands the area available for your source-code windows.

Do the same thing with the Output and Task List window tabs at the bottom of your screen and you will free up even more space for your editing windows.

To help you minimize screen clutter, you can set any of the tool windows to Auto Hide. If you look on the left side of your screen, you will see that the Server Explorer and Toolbox windows have the Auto Hide property set by default. You cannot, however, set the source code windows to Auto Hide.

You should spend some time experimenting with the Auto Hide and Dockable properties. It is like putting your tools away when you are not using them, leaving your workbench clear for the work at hand. When I first started Visual Studio.NET, I felt cramped by all of the tool windows encroaching on the screen. However, setting most of the windows to Auto Hide has greatly reduced the claustrophobic feeling.

Exploring Visual Studio Menus and Toolbars

Toolbars and menus are as plentiful in Visual Studio as tool windows, perhaps even more so. Just when you think you have seen all of the possible menus, another will pop up somewhere.

The menus along the top of the Visual Studio are on the *main menu bar*. You cannot hide the main menu bar, nor can you move it, although you can move toolbars and place them above the main menu bar. As you will learn as you progress through this book, Visual Studio may remove some of the menus and add others according to the task at hand. For example, if you are working on a Windows *form*, the Visual Studio will add Format menu to the main menu bar.

The other menus that appear at various places around the Visual Studio are *context menus*. The meaning of these menus depends upon the context in which you summon them. Throughout this book you will be asked to "right-click" your mouse on an object such as an item in the Solution Explorer, then select an item from the menu that pops up—the context menu.

To get an idea of how the context menus differ, even between different lines in a window, open the *Hello* project from Lesson 2. Move the mouse cursor to the Solution Explorer and right-click your mouse on the line that reads "Solution 'Hello' (1 project)." Examine the contents of the menu, then close it without selecting any of the items (click on a blank area of the Visual Studio frame to hide a context menu).

Now move to the next item, the line that reads just "Hello," and right-click on it. Examine the contents of this menu. The two menus are shown side by side in Figure 3.5.

Figure 3.5. The context menus for the top two lines in the Solution Explorer window show some similarities, but many of the options are different for each menu.

Notice that some of the items on the two menus are the same. You can "build" a solution or a project, so that option appears on both. However, a solution may have multiple projects, so the Rebuild All item appears on the Solution menu, but would be meaningless on the Project menu.

In addition to the menus, Visual Studio is loaded with toolbars—28 in all, and you may create your own toolbars to fit your own needs.

To display a list of the toolbars, right-click your mouse in an open area of the Visual Studio window frame. The context menu that appears is a list of the available toolbars. If a toolbar is visible, it has a check mark next to it. Note the Customize item at the bottom; you will return to it shortly.

You may move a toolbar to any place on the Visual Studio frame. If you "dock" the toolbar to the left or right side, the toolbar will orient itself vertically. To move a toolbar, move the mouse to the far left side of the toolbar. Each toolbar has a vertical mark at the far left side to mark the grabber, as shown on the Text Editing toolbar in Figure 3.6. The mouse cursor should turn into a set of crossed arrows. Click and hold the mouse and move the toolbar to where you want to place it.

Figure 3.6. The vertical mark at the far left of the toolbar is the "grabber." Use the grabber to move the toolbar. At the far right is a down arrow, which you use to add or remove buttons.

Again referring to Figure 3.6, each toolbar has a down arrow at the far right side. Click the mouse on the arrow to reveal a menu to add or remove buttons. Click your mouse on this item to list the buttons on the toolbar. Each button listed has a check mark next to it if it is visible in the toolbar. Click on a button to toggle it on or off. At the bottom of the list is an item to restore the toolbar to its original state.

Return to the down arrow button and click it again, then click the Add or Remove Buttons item again. You will see a Customize option at the bottom of the menu for each menu. This is the same Customize item you saw when you displayed the list of toolbars a couple of paragraphs ago. Click on this item to create your own toolbars. The Customize dialog box has context-sensitive help; right-click on any item and click the What's This? item to summon help.

Using the Solution Explorer

The Solution Explorer window is the primary tool through which you will manage your workspace, or "solution." The Solution Explorer for the *Hello* project is shown in Figure 3.7. The *Hello* project's Solution Explorer is relatively simple; a project normally would contain more references and probably several source-code files.

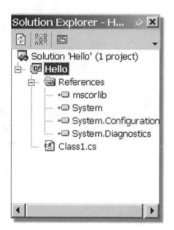

Figure 3.7. The Solution Explorer is a navigation aid to your solution and to the projects it contains.

The first thing you should note about the Solution Explorer window is that it has a small toolbar at the top. The toolbar makes the window a navigation aid to your solution and to the projects within it. As you select different items within the Solution Explorer, the toolbar items will change to reflect the actions that are available for that item. If you click your mouse on the first item in the window, only a Properties button will appear, but if you click on a source-code file, such as *Class1.cs*, up to four buttons will be available. Clicking on the last button opens the source-code file in an editing window.

Right-clicking your mouse on an item in the Solution Explorer also reveals a context menu, the contents of which will vary according to the type of the item selected. The menu for a project item, for example, might contain items to start the project in the debugger, but the menu for a source-code file would not. Instead, a source-code item's context menu would contain items related to editing, such as opening the code in an editing window or cutting, copying and pasting text.

Using Class View

As you will learn in the next few lessons, everything in C# is built around classes. A class is the primary mechanism through which C# implements *object-oriented programming*. All functions, objects, and variables must be members of a class.

The Class View window summarizes the classes in a project and lists all the functions, objects and variables that are members of the classes. Like the Solution Explorer, Class View is a navigation tool that gives you the ability to move to specific places in your code. In addition, Class View is a primary mechanism through which you may maintain classes in your program. The Class View window for the *Hello* project is shown in Figure 3.8.

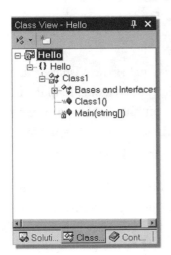

Figure 3.8. The Class View window contains a summary of classes in your project. Class View also will help you to maintain your classes.

Like the Solution Explorer, the Class View window has its own toolbar. However, the Class View toolbar does not change when you select different items. The first button on the toolbar is the Sort button. Clicking your mouse on the Sort button sorts the items in the classes according to a predefined criterion. If you click on the arrow next to the Sort button, you will get a list of the criteria used for sorting the class items. The second button adds a new folder to the Class View window.

Using the Class View, you may add *methods* and *properties* to your class. A method is a function member that performs a specific task, and a property is a variable that you may access and set using *Get()* and *Set()* functions. In addition, you may add *fields*—variables—and *indexers* to your class. You will learn more about these terms in the next few lessons.

The Class View window helps you to navigate your code. If you double-click your mouse on an item in a class, the Visual Studio will open the source file and move to the definition of the item. Using the *Hello* Class View as an example, close the source file window (right-click on the source code window tab and select Close from the pop-up menu). Now, double-click on the *Main()* item in the Class View list. Visual Studio will open the *Class1.cs* file and move the caret to the *Main()* function.

Getting Help

The help system for Visual Studio.NET is the Microsoft Developers Network (MSDN), which you installed when you installed Visual Studio and C#.

You may configure the help display by selecting Options from the Tools menu. When the Options dialog box appears, click on the Environment item to expand the list, then select Help. The Internal Help option makes the help window appear in the text editing area. If you select internal help, the Visual Studio also will add a Search window on the right side of your screen (near the default location for the Solution Explorer) in which you may enter topics for which you need help. All of this tends to make the Visual Studio seem even more crowded.

External Help makes the MSDN run as a separate program, which you may minimize, maximize and close independently of the Visual Studio. If you change this option, you must exit and restart the Visual Studio for the change to take effect.

Regardless of whether you select Internal or External help, the Visual Studio maintains a Dynamic Help window. The default location for Dynamic Help is just below the default location for the Solution Explorer on the right side of your screen. It shares that location with the Properties window. Select the Dynamic Help window by clicking on the Dynamic Help tab at the bottom.

Dynamic Help is *context tracking* help; the window contents will change according to the task you are performing. With the *Hello* project open, select the Solution Explorer. Click your mouse on each item in the Solution Explorer and watch the contents of the Dynamic Help window change with each selection. Now move to the *Class1.cs* source code window and click on the *public* keyword in the *Class1* definition. Examine the contents of the Dynamic Help window. Now click on the *static* keyword in the *Main()* function definition and re-examine the Dynamic Help window. The contents will change so that you may look up help for the current task.

Using the Find In Files Command

The Visual Studio has the standard find and replace commands that users have come to expect in any Windows program. You can access these commands by selecting the Edit menu, then the Find and Replace item.

In addition to the standard Find and Replace commands, Visual C# also has a command that will search *all* the files in your project and another that will perform a text replacement in *all* the project files. Select the Edit menu and select the Find and Replace item. From the drop-down menu select Find In Files. You will get a dialog box like the one shown in Figure 3.9.

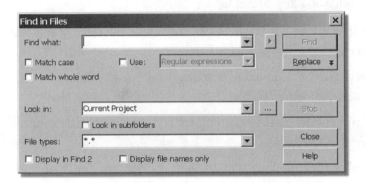

Figure 3.9. The Find In Files dialog box allows you to select groups of files to search.

By default, the Find In Files command will search all the files in your project, but you may use the Look In field to limit the search to certain files. Clicking your mouse on the Replace button to the right changes the dialog box into a Replace In Files dialog box. However, there does not seem to be any way to reverse the change.

Using the *Hello* project again, summon the Find In Files dialog box as just described. In the Find What field, type **Class1**. Click on the Find button. When the command ends, you will notice the Visual Studio has added another tab, Find Results 1, at the bottom of the screen in the area shared by the Output and Task List windows. The results of the search will appear in this window. Double-click your mouse on a line to move to the file containing the entry.

Summon the dialog box again. Notice at the bottom that there is a box labeled Display in Find 2. Check this box, then type **Main** in the Find What field. Click on the Find button. When the command completes, notice that the Visual Studio has added yet another tab, Find Results 2, at the bottom of your screen. By using two panels, Visual Studio gives you the ability to perform a second search without overwriting and destroying the results of the first search.

WHAT YOU MUST KNOW

In this lesson, you have learned how to use many of the features and tool windows in Visual Studio.NET. You learned how to set options for your projects and how to use the Solution Explorer and Class View windows. You also learned the basics of finding help for the tasks you are performing and how the Visual Studio Dynamic Help window tracks the work you are performing and adjusts to display help topics available for your current task. Finally, you learned how to use the Find In Files command to search the source-code files in your project. In Lesson 4, "Understanding a Few Key C# Basics," you will begin to explore the C# programming language. Before you continue with Lesson 4, however, make sure you have learned the following key concepts:

- The Visual Studio Options dialog box is where you customize the display and set options for your project.

- To create a new project, select the programming language and a template for the project by using the New Project dialog box.

- Visual Studio contains many tool windows, toolbars and menus at various locations around the main window frame.

- You may hide or display tool windows and toolbars as needed. You also may make tool windows display when needed using the Auto Hide property. Displaying only tool windows and toolbars that you need removes much of the "clutter" from the Visual Studio window.

- You display context menus by right-clicking the mouse on an item in one of the tool windows. A context menu contains items and commands available for the selected item.

- The Microsoft Developers Network is the help system for the Visual Studio. In addition to the Help menu, Visual Studio contains a Dynamic Help window that tracks your mouse and caret movements to display available help topics.

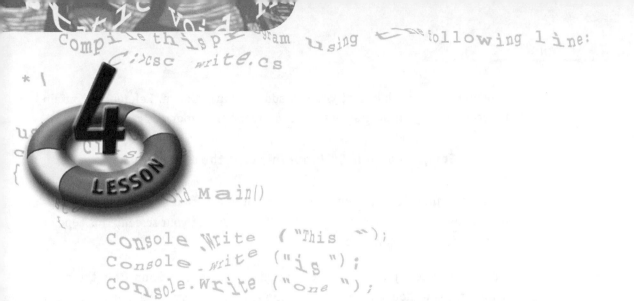

```
Compile this program using the following line:
C:>csc write.cs

* \

us

{

        oid Main()

    {
        Console.Write ("This ");
        Console.Write ("is ");
        Console.Write ("one ");
```

USING C# FROM THE COMMAND LINE

*t*he primary purpose of the Visual Studio is to help you to write programs for Windows, or components for Windows programs. Windows programs typically take some degree of planning and setup to design and create the windows and forms that a Windows program needs. Many times, however, you will want to bypass that process and write a simple program that runs from the Windows command line. Writing short, command-line programs also is a good way to learn the syntax of C# and to test program ideas. In this lesson, you will learn how to write and compile command-line programs, and how to run your programs from the Windows command line. You also will learn how your program may access and use Windows objects from the command line. By the time you finish this lesson, you will understand the following key concepts:

- You may invoke the C# compiler from the Windows command line to compile a program without the Visual Studio.
- A *function* is a named block of code that performs a specific purpose. In C#, you write all of your program code in functions.
- Every C# program must have a function named *Main()*. When you run your program, Windows starts your program by calling *Main()*.
- The *Console* class contains functions to write text to your screen and read text that you type from the keyboard.
- A string is a group of characters in which you may store text. You use strings to convert information in your program to text that you can read on the screen.

- A *reference* is a module that you can add to your program. References contain functions that your program may call to extend the program's capabilities.

Compiling Your C# Program from the Command Line

In the beginning lessons, you have written programs that you run from the MS-DOS command line and write to the MS-DOS window of your screen. This type of program is a console program.

When you write a program that you intend to compile and run from the command line, you should use an editing program that can save the files as ordinary text files. A good editor for this purpose is NotePad, which Windows provides. Word processing programs such as Microsoft Word or Corel WordPerfect normally save your text along with formatting information, which the C# compiler does not recognize.

To run NotePad and create a C# program file, first start a command-line window. Visual Studio has its own command window that you should use for developing programs instead of the Windows command prompt. The Visual Studio command window sets up the environment variables that you will need to use the C# compiler. To start the Visual Studio command window, select the Start menu, then Programs. Locate the Microsoft Visual Studio.NET 7.0 item and select it. From the submenu, Visual Studio.NET tools. Finally, select Visual Studio.NET Command Prompt from the next menu that pops up.

After the command prompt window displays, change to the directory where you will keep your C# program files. C# source files normally have an extension of *.cs*, so type the following command line to create a C# program:

```
C:> notepad myprog.cs   <Enter>
```

If *myprog.cs* does not exist, NotePad will ask you if you want to create the file. Answer Yes and then enter the following text (the text in a C# program file is case-sensitive, so enter the text exactly as shown):

```
using System;
class clsFirst
{
    static void Main ()
    {
        Console.WriteLine ("This program was written");
        Console.WriteLine ("with NotePad and compiled");
        Console.WriteLine ("from the command line.");
    }
}
```

Save the file and exit NotePad. You now should have a file in your directory called *myprog.cs*. The C# compiler will read this file and create an executable file by the same name but with an extension of *.exe*. Type the following command line to compile *myprog.cs*:

```
C:> csc myprog.cs   <Enter>
```

If you made any mistakes typing in your program, the compiler will display any error messages after these lines. If you did not make any mistakes, the compiler will create *myprog.exe* and the command prompt will reappear.

Running Your C# Program

If you are familiar with the MS-DOS command line, you can skip this section. However, if you have never started a program from MS-DOS (and there are many people who have not), there are some basics you should understand.

When you compiled your program in the last section, the compiler created an *executable* program file named *myprog.exe*. The *.exe* part of the name is the file *extension*, and tells the Windows command processor that this is an executable file. The command processor knows about certain file extensions that are executable, such as *.com*, *.exe*, *.bat* and *.cmd*. If you type the name of a program without an extension, the command processor will look for a file with one of

these extensions. If the command processor finds a match, it will load the file and transfer control to the program file.

To run your program file, *myprog.exe*, then, you need only type *myprog* at the command prompt and press the **Enter** key:

```
C:> myprog   <Enter>
```

You should see the following output:

```
This program was written
with NotePad and compiled
from the command line.
```

Later versions of Windows such as Windows 2000 do not care what the extension is as long as the executable file has an extension. You could call your program *myprog.stuff* and you could run it by typing the full name of the file on the command line:

```
C:> myprog.stuff   <Enter>
```

You cannot name an executable file without an extension, however, because the command processor will try to add one of its known extensions to the file when you type its name. For example, if you name your program file simply *myprog* and type its name at the command prompt, the command process will attach *.exe* to the name and look for the file, then it will attach *.com* to the name and look for the file again until it uses up all of its known extensions. Finally, the command processor will tell you that it cannot find the program file.

Understanding the *Main()* Function

As you will learn, functions are the basic building blocks of your C# program. You can declare *variables* and define *objects* outside of a function, but any code that you write in a C# program must be in a function.

A function is a named block of related code that performs a specific purpose. By naming these blocks of code, you enable your program to execute the code the functions contain by *calling* the functions. Your program performs a *function call* when it executes a statement in which you have written the name of the function. When your program code enters a function, it executes the statements in the function one after another, beginning with the first statement in the function. Your program continues executing the statements until it gets to the last statement or encounters a *return* instruction. The function will then exit—or return—and the program control goes back to the statement after the line containing the function call.

Often in C# documentation, you will see functions referred to as *methods*. Technically, in object-oriented programming, a method is a function that an object exposes to perform a specific request. However, as you will learn, in C# all functions must be members of a *class*, and thus all functions qualify as methods. Do not be overly concerned with the distinction; just be aware that in the C# literature you may see the words *function* and *method* used almost interchangeably.

A C# program may contain as many functions as you need, and you may give a function virtually any name you want. However, every C# program must have at least one function named *Main()*. Notice that the name begins with a capital M. If you have programmed in C or C++, you have learned the only required function is *main()* with a lower case m. In C#, the *Main()* function serves the same purpose as the C and C++ *main()* function.

When you run a C# program, the operating system (Windows) passes control to the first statement in *Main()*. The program then executes the statements in *Main()* one after another. When your program performs the last statement in *Main()*, it returns control to the operating system and your program ends. Between the first and last statements, however, you may create objects and call other functions from within *Main()*.

A function may or may not return a value to the statement that called the function. If the function does not return a value, the function's return type is *void*. To set up *Main()* so that it does not return a value, you would define it as follows:

```
static void Main()
{
    //  Program statements go here.
}
```

As you will learn in later lessons, a function may return a value of a specific *data type*. You may set up the *Main()*function to return a single number value, an integer, by replacing the *void* type with *int*. In this case, you must terminate the *Main()* function with a *return* statement:

```
static int Main()
{
    //  Program statements go here.
    return (0);
}
```

The *static* keyword gives the *Main()* function a special status. In C#, all functions—including *Main()*—must be members of a class. However, you cannot call a member function until you create a class object. This leads to a Catch 22 situation. The *Main()* function is where you create your initial objects, but how do you create an object of the class that contains *Main()* until you call the *Main()* function?

C# gets around this problem by requiring that you declare the *Main()* function as *static*. Your program may call functions that you declare as *static* at any time, even before you create an *instance* of the class that contains the *static* function. (An *instance* of a class occurs when you declare an *object* of the class.) Functions declared *static* have special properties, and special limitations, as you will learn in Lesson 7, "Getting Started with Classes."

The following program, *Simple.cs*, declares a class to hold the *Main()* function, then uses *Main()* to write to your console:

```
/*
    simple.cs. A simple command-line program that writes
    a single line to the console.

    Compile this program using the following line:
        C:> csc simple.cs
 */

using System;
class clsSimple
{
    static void Main()
    {
        Console.Write("This line writes to your screen\n");
    }
}
```

You also may pass arguments to the *Main()* function when you run your program. Windows sends the arguments to *Main()* as a *string array* which you access using *index* values. You will learn about arrays and array indexes in Lesson 16, "Using Array Variables to Store Multiple Values." For now, you should understand that an array contains multiple values, and you access the first value in the array using an index value of 0:

```
/*
    simplArg.cs. A simple command-line program that accepts
    command-line arguments and writes them to the console.

    Compile this program using the following line:
        C:> csc SimplArg.cs
 */

using System;
class clsSimple
{
```

```
static void Main(string[] args)
{
    int count = args.Length;
    if (count == 0)
    {
        Console.Write("You didn't enter any " +
                      "arguments\n");
        return;
    }
    int i = 0;
    while (i < count)
    {
        Console.Write("Argument {0} is {1}\n",
                      i, args[i]);
        i = i + 1;
    }
}
}
```

Run this program by typing the program name along with some numbers on the command line, similar to the following:

```
C:> simplarg 1 2 3 4   <Enter>
```

The program will list the arguments one by one, starting with the first argument (argument 0):

```
Argument 0 is 1
Argument 1 is 2
Argument 2 is 3
Argument 3 is 4
```

You should notice two new constructions in this sample. In the first, you test whether you typed any arguments when you ran the program by using the *if* statement. This is an example of a *conditional* statement, and it gives your program the ability to make decisions. You will learn more about conditional statements in Lesson 20, "Making Decisions Within a C# Program." Notice especially the *return* statement. This causes *Main()* to stop executing any further statements and to exit. When *Main()* returns, your program ends.

The second construction is the *while* statement. This statement causes your program to repeat the following statement while a given condition is true. This is an example of a *loop*, which you will study in Lesson 21, "Repeating Statements Within a C# Program."

Performing Basic Output: The Console Class

When you write a command-line program, from time to time your programs must write to the screen or read characters from the keyboard. In the C programming language, every program inherited a set of *streams* from the operating system that enabled the programs to read from the keyboard and write to the screen.

C# also uses streams, but it handles them differently. C# isolates your program from the operating system by placing the keyboard and screen streams into the *Console* class. C# then uses member functions in the *Console* class to manage your program's access to the keyboard and the screen.

Streams Handle Data Flow

A stream is a sequence of data flowing from one part of your computer to another. For example, your program cannot directly access the keyboard because doing so might interfere with another program's attempt to read keyboard characters. Instead, Windows handles keyboard input and directs the characters you type to the proper program using a stream. In C, the keyboard stream is called stdin.

Similarly, you cannot write directly to the screen because your program only has access to the part of the screen its window occupies. Windows takes the output from your program and displays it in that window using another stream, stdout.

From a practical standpoint, both ends of the streams are processes. For the keyboard, the stream connects with a driver *program, which places the keystrokes into the stream. The characters resulting from the keystrokes then appear at the other end of the stream, which is your program. Windows handles the details of directing the character through the stream.*

An advantage to using streams is that a stream may be redirected. *Your program can move one end of the stream from one device to another. When you issue an MS-DOS command such as* DIR > SomeFile.txt *the operating system will redirect the DIR command's output stream* (stdout) *from the screen display to a file named* SomeFile.txt.

Up to now, you have been using the *Write()* function in the *Console* class to write to the screen. *Write()* directs the information exactly as you specify to the standard output stream. It does not write a line-ending sequence (a carriage return and line feed) to the output stream, so you may use individual calls to *Write()* to build a single output line. Another function, *WriteLine()*, has much the same purpose except that it writes a line terminator to the output stream. The following program, *Write.cs*, shows how you can build a single line from several calls to *Write()*, but *WriteLine()* always ends the line:

```
/*
    write.cs. A simple command-line program that writes
    to the console using Console.Write() and Console.WriteLine().

    Compile this program using the following line:
        C:> csc write.cs
 */

using System;
class clsSimple
{
    static void Main()
    {
        Console.Write ("This ");
        Console.Write ("is ");
        Console.Write ("one ");
        Console.Write ("line ");
        Console.Write ("of ");
        Console.Write ("text.\n");
```

```
    Console.WriteLine ("This ");
    Console.WriteLine ("is ");
    Console.WriteLine ("several ");
    Console.WriteLine ("lines ");
    Console.WriteLine ("of ");
    Console.WriteLine ("text.");
  }
}
```

Notice that in the last call to *Console.Write()*, the text includes a backslash followed by a lower case *n*. This is the C# representation for a *newline* character, which actually is a combination of the carriage return and line feed on Windows. Because the newline character may be different on other operating systems, and Microsoft hopes to make C# available on other systems, the "\n" sequence always will produce the proper line ending sequence.

The *Console* class also contains two functions that read from the keyboard, or the standard input device. The first, *Read()*, accepts only one character at a time. The second function, *ReadLine()*, accepts characters until you enter a newline character.

For compatibility with other systems, C# defines the ending characters as a carriage return (the line-ending sequence on MacIntosh computers), a line feed (the line-ending sequence on UNIX systems) or both a carriage return and line feed (Windows).

In the following program, *Read.cs*, notice that the *Read()* function will accept only one character, regardless of how many characters you type. The *ReadLine()* function on the other hand will accept all the characters you type:

```
/*
    read.cs. A simple command-line program that reads from
    the console using Console.Read() and Console.ReadLine().

    Compile this program using the following line:
        C:> csc read.cs
*/
```

```
using System;
class clsSimple
{
    static void Main()
    {
        Console.WriteLine ("Using the Read() function");
        Console.Write("Type one or more characters: ");
        int arg = Console.Read();
        Console.WriteLine (Convert.ToChar(arg));

        Console.WriteLine ("\nUsing the ReadLine function");
        Console.Write ("Type one or more characters: ");
        string str = Console.ReadLine ();
        Console.WriteLine (str);
    }
}
```

You are about to encounter your first "gotcha" in a C# program. Compile and run the program from the command line. When you are prompted to type one or more characters, press just one character key on your keyboard. Notice that the *Read()* function does not return when you enter one character. Instead, you have to press the **Enter** key.

The "gotcha" comes from the fact that you have just added another keystroke to the stream, the newline. *Read()*, however, accepts only one character, leaving the newline in the stream. When you subsequently call *ReadLine()*, it accepts the newline as the end of the input and returns immediately. Although *ReadLine()* strips the newline off, it still reads the remaining characters in the stream.

Actually, this is not bad behavior on the part of *Read()* and *ReadLine()*. If you redirect the stream to a file, that is how you would want to read the file. It is, however, bad news when you are trying to read the keyboard character by character (such as in a communications or network chat program). In the current version of C#, there does not appear to be any way around this. *Read()* is supposed to return a –1 when no additional characters are available.

To get around this, make sure the input stream is empty by calling the *ReadLine()* function after you accept a single character, then ignore the input.

To make your program independent of a particular operating system, you cannot simply continue to read a single character using *Read()* until you reach a new-line; you do not know whether the newline is a carriage return (on a Mac) or a line feed (on UNIX), or both (on Windows).

```
/*
    read2.cs. A simple command-line program that reads from
    the console using Console.Read() and Console.ReadLine().

    Compile this program using the following line:
        C:> csc read2.cs
 */

using System;
class clsSimple
{
    static void Main()
    {
        Console.WriteLine ("Using the Read() function");
        Console.Write("Type one or more characters: ");
        int arg = Console.Read();
        Console.WriteLine (Convert.ToChar(arg));
//
//  The following line flushes the standard input stream
//  and ignores the result.
//
        Console.ReadLine();

        Console.WriteLine ("\nUsing the ReadLine function");
        Console.Write ("Type one or more characters: ");
        string str = Console.ReadLine ();
        Console.WriteLine (str);
    }
}
```

Before leaving this section, you should notice that the value returned by *Read()* is an integer value rather than a character. If you try to write the value directly to the screen, the *Write()* function will print the *value* of the integer rather than the character itself. Try substituting the following line after the call to *Console.Read()* to see what happens:

```
Console.WriteLine (arg);
```

C# treats everything as an object, even numbers, and there are some built-in conversion functions that you may use. To convert a value from one data type to another, you may use one of the functions in the Convert class. One of these functions is ToChar() to convert an integer value to a character. For example, writing Convert.ToChar(65) would convert the integer value 65 to the character A.

Using Strings and String Literals: "Hello, C# world!"

In the previous section, you read the input from the keyboard in two ways. First, you read a single character and saved the result in a variable, *arg*. The variable can hold a single value at any given time. Then, you used the *ReadLine()* function to read all of the characters from the keyboard into a *string*.

When you put two or more characters together, you are building a string. Although C# treats a string as a single variable, a string is an example of an *array*, a special variable type that can hold more than one value at any given time.

Microsoft designed C# to be an international computer language. In C#, a character value is a 16-bit value, which is two bytes of memory. A single character, then, can hold a value up to 65,537 and can represent the many characters that are used in various languages around the world. When you declare a *string*, you set aside enough memory to hold one two-byte value for each character in the string.

Arrays Store Multiple Values

Normally, a variable can hold only one value at a time. Sometimes, however, you must store more than one value in a variable. In a string, for example, you can store several characters (such as a user's name or a file name) that you can access using just one variable name.

*You can declare a variable as an array variable; the C# compiler will set aside enough memory to hold more than one value. You specify an array by writing a set of double brackets after the data type, declare the variable, and then create the array using the **new** operator. You specify how much memory to set aside by writing the number of values you want to store inside square brackets when you declare the array variable: The following declaration sets aside enough memory to hold 10 integers:*

```
int [] MyInts = new int[10];
```

*You then may access individual values by telling the compiler the **index** of the value inside square brackets. The index is 0-based, so the first element is 0, the second is one, and so on. The following statement accesses the fourth element in the array and sets the variable **OneInt** to that value:*

```
int OneInt = MyInts[3];
```

*When you declare a string, you actually are declaring an array of characters. The **System.String** class handles the details of declaring the array.*

You will learn more about arrays in Lesson 16, "Using Array Variables to Store Multiple Values."

A C# string appears to be a basic data type and you can perform certain operations on string, such as adding two or more strings and testing whether two strings are identical. However, the *string* keyword actually refers to a *String* class, which defines how the characters are stored and what operations you may perform on a string.

At this writing, the *String* class appears to be evolving still (C# is a new computer language), and some books may differ on how C# handles strings. Be sure to read the *string* help page that comes with your version of Visual Studio.

To declare a string, write the keyword *string* followed by the name of the variable and a semicolon:

```
string MyString;
```

You also may assign text to a string when you write the declaration by writing an equal sign and the text. In the following statement, *MyString* is a *string variable* and the text inside the double quote marks is a *string literal*:

```
string MyString = "This is a string.";
```

You may read in the documentation and in other books that you may declare a string using the *new* operator. There are two things to remember about this method. First, you *must* initialize the string by placing the text inside a set of parentheses:

```
string MyString = new string ("This is a string.");
```

Second, you should be aware that this method of declaring a string may disappear in future releases of the Visual Studio. Although the compiler will not issue an error, you will get a warning that this method is "obsolete":

```
Read.cs(14,22): warning CS0618:
'string.String(string)' is obsolete: 'Use
String.Copy if you really need a copy.'
```

In the last section, you might have noticed the "\n" sequence to create the newline character. When you assign text to a *string*, the C# compiler searches your text for a backslash ("\") character. The compiler treats the backslash as an *escape* character, and translates the *next* character to another, predefined value. The combination of a backslash and another character is an *escape sequence*. C# defines several escape sequences as shown in Table 4.1.

Escape Sequence	Meaning
\a	Bell character
\b	Backspace character
\f	Formfeed character
\n	Newline
\r	Carriage return (no line feed)
\t	Horizontal tab
\v	Vertical tab
\"	Double quote character
\\	Backslash character

Table 4.1 Escape sequences recognized by the C# compiler.

Some of the escape sequences are historical. For example, the \b sequence rings the bell on the Teletype machines that once were used as system console devices; now it generates a sound on your computer's speaker. If you use a backslash character followed by any character not in the list, the C# compiler will generate an "Unrecognized escape sequence" error.

You should remember that although you type two characters—the backslash and the escape character—the compiler converts them into a single character. When you count the number of characters in your string or access them using an index, count an escape sequence as only one character.

To enter a backslash, you must type it twice. This is important if you are typing the name of a Windows directory. The following path in a C# string actually produces "C:>\WINNT\System32":

```
string path = "C:>\\WINNT\\System32";
```

C# also provides a method of assigning text to a string without processing escape sequences. To do this, you precede the string literal (the text) with the @ character:

```
string path = @"C:>\WINNT\System32";
```

When you use the @ character before text, the C# compiler will *not* process *any* escape sequences. If you add an "\n" to generate a newline, the string will contain the backslash and the *n*, but no newline character. This also means that you may not embed a double quote mark in a string, even by escaping the character. Try the following three statements:

```
string str1 = "The word is \"stuff\""
string str1 = "The word is "stuff"";
string str2 = @"The word is \"stuff\"";
```

The first statement will compile properly, and if you print out the *str1* variable, you will see the desired output:

```
The word is "stuff"
```

However, the second and third lines both will cause the compiler to generate errors and write them to your screen.

Writing a string literal by prefixing it with the @ character is specific to C# and you should avoid it if possible. If you later go on to learn C or C++, or to program on systems other than Windows, the difference will only add confusion.

You can add strings using the + symbol. When you do this, C# will add the second string to the end of the first string, producing just one string. In the following, the two statements yield the same string:

```
string str1 = "This is an example" + "of adding strings";
string str2 = "This is an example of adding strings";
```

You can use this fact to make several output lines in your program appear as one line to an output function:

```
Console.WriteLine ("By adding strings, you " +
                   "can make several lines " +
                   "appear as one");
```

If you add this code to one of your programs, it will produce the following output as a single line:

```
By adding strings, you can make several lines appear as one.
```

Because of the constraints of printing, not all the text that will fit on your screen will fit on the width of a page in this book. The examples in this book frequently will use this technique to show long code and output lines.

Writing a C# Program for Windows

Normally when you write a command line program, you will write your message to the user on the console screen using the *Console.Write()* or *Console.WriteLine()* functions. For input, you normally will use the *Console.Read()* function.

By adding a statement that you are using the *System.Windows.Forms* namespace to your program, however, you can take advantage of some Windows objects such as the message box. Later, when you learn to use *forms* in your program, you can use forms to accept input from the user.

Even though most of your programs so far have been command line-based, you have had access to classes that implement Windows objects by using the *System*

namespace in your program. Adding the line that you are using the *System.Windows.Forms* namespace makes access to those object easier.

The code to implement the Windows objects is part of the *mscorlib.dll* assembly. The compiler automatically adds a reference to *mscorlib.dll* to your program when you compile the program. You will learn how to write your own modules and add them as references to another program in Lesson 13, "Using References and Assemblies."

Enter the following short program, *ref.cs*, using NotePad as described in the first section of this lesson. Note that you now have two statements that you are using other namespaces:

```
//
//   Ref.cs — Demonstrates using the System.Windows.Forms
//             namespace to access the MessageBox class
//
//             Compile this program with the following
//             command line:
//                  csc Ref.cs
//
using System;
using System.Windows.Forms;

public class clsHello
{
    public static void Main()
    {
        MessageBox.Show ("This command line program " +
                         "uses a message box",
                         "Using Windows Objects");
    }
}
```

Compile the program with the following command line:

```
C:>csc ref.cs
```

When you compile and run *Ref.cs*, the .NET runtime loads the *mscorlib.dll* assembly into memory and you automatically have access to the Windows objects. You will see the message box shown in Figure 4.1.

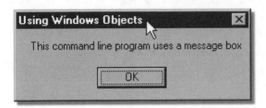

Figure 4.1. By adding the System.Windows.Forms namespace to your program, you can create and use Windows objects from the command line.

The *System.Windows.Forms* namespace also contains classes to implement the Windows *common controls* and *common dialogs*. You will learn how to use the controls in Lesson 28, "Using Windows Controls," and the common dialogs in Lesson 30, "Using the Common Dialogs in a C# Program."

WHAT YOU MUST KNOW

In this lesson, you learned how to use the C# compiler from the command line. Writing command-line programs and compiling them from the command window is a good way to learn the syntax and elements of C# because you can concentrate on the language essentials rather than having to manage the windows in the Visual Studio. You learned the basics of using *strings* in C#. Strings are important because you can use them to present program information in a form that your users can read. In addition, you learned how to add references to your program. In this lesson, you added references that enabled you to create and use Windows objects, but later you will create your own components using C# and add them to other programs as references. In Lesson 5, "Understanding a Few Key C# Basics," you will begin to examine the basics of object-oriented programming. Before you continue with Lesson 5, however, make sure you have learned the following key concepts:

- You can write programs for the command line without using the Visual Studio environment. To do this, you write your programs using a plain text editor such as *NotePad*.

- Statements that your program executes must be in a *function*. A function is a named block of code in a C# program.

- Your C# program must contain a function named *Main()*. When you run your program, Windows transfers control to the first statement in *Main()*.

- You can write to the console screen and read from the keyboard using functions in the *Console* class.

- A *string* contains text. A *string variable* is a variable that can hold many characters. You use strings to hold text that you read from the keyboard or to display to the user.

- Your program may access other code modules by adding the modules as *references* when you compile your program. When you add a reference, your program may access and use functions in the reference.

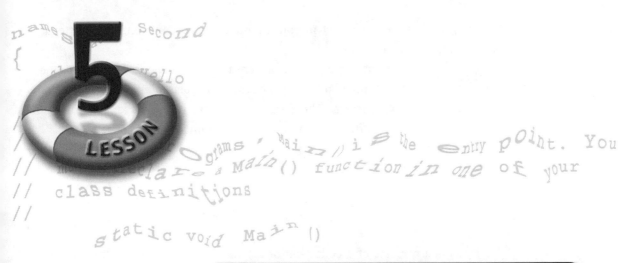

UNDERSTANDING A FEW KEY C# BASICS

*n*ow that you have learned how to use the various tool windows and toolbars around the Visual Studio, and how to run the C# compiler from the command line, it is time get into the basics of programming in C#. As with any computer programming language, C# has a set of rules—a *syntax*—that you must follow to write instructions. Up to now, you have been dealing with very simple programs. Soon, however, you will begin adding *variables* and *functions* to your program. To do this, you must understand the rules and the conventions used in C#. In this lesson, you will begin learning about C# syntax. You will learn how to add *white space* to your code to make it more readable, and how to add *comments* to remind yourself and to inform other programmers why you wrote code in a certain way, why you declared a variable of specific type and to describe the purpose and result of a function. By the time you finish this lesson, you will understand the following key concepts:

- A C# program consists of *tokens* and *white space*. Tokens are the keywords, operators, variable and function names and symbols you use in a language. You use white space to separate the tokens.
- A *statement* is a single instruction that you write in your program. In C#, a statement ends with a semicolon.
- A statement may span several lines. You may use this fact to make your program code more *readable*.

- You may include narrative text in your programs using comments. The C# compiler ignores commented text.
- The C# language contains two types of comment markers: single-line and multi-line.
- A *namespace* is a container for the objects such as classes and structures that you define in a C# program. The names of objects in a namespace must be unique, but names in different namespaces may duplicate on another.
- When you make changes to your C# program, you must recompile your program for the changes to take effect.

Tokens and White Space

A C# program consists of a collection of *tokens* and *white space*. White space is any character that only moves the caret on your screen or changes the position of the character on your printer. White space characters do not display anything on your screen or printer.

The space character is the basic example of a white space character. It causes the caret to move one space to the right but does not display anything on your screen. The tab character also is a white space character. It moves the caret to the next preset tab position, but does not display on your screen or printer.

Other white space characters are the carriage return and line feed characters, which end the current line you are typing and move the caret to the beginning of the next line. The carriage return and line feed characters normally appear together, and DOS and Windows programmers often refer to them as a single character, the *newline*. You type a newline by using the **Enter** key.

Anything that is not white space is a *token*. A token may be a single character, as in the '+' operator, or it may be two or more characters, as in the name of a variable such as *MyVar* or the name of a *class* or any other object you define in your C# program.

C# uses white space to separate tokens, and you may have as many white space characters between tokens as you need, which is handy for aligning blocks of code or variable declarations. However, tokens may not themselves contain white space characters. For example, the following is a valid C# variable declaration:

```
int    MyVar;
```

This line declares a single integer variable using the int keyword and the characters *MyVar* as a token to represent the variable. If, however, you write something like the following, the compiler would find two variable tokens rather than one and generate an error:

```
int    My Var;
```

The C# compiler ignores multiple white space characters. Because the newline is a white space character, this means you can use a combination of spaces, tabs and a newline to write a single statement on more than one line, which helps to make your code more readable. Rather than having one long statement across the line, you can break the line at any point and simply continue on the next line. Programmers often use this technique to make their code easier to read. Examine the following loop statement (you do not need to understand how the loop works yet):

```
for (int x = 0; x < 10; x = x + 1) y = x * x;
```

The statement could be rewritten as follows without changing the way it works:

```
for (int x = 0; x < 10; x = x + 1)
   y = x * x;
```

Now the loop statement is on one line and the dependent statement on another. When reading the code, the indent immediately tells you that it is dependent upon the preceding line. Because C# ignores extra white space between tokens, you could carry this even further and write the statement as follows:

```
for (int x = 0;
     x < 10;
     x = x + 1)
        y = x * x;
```

This is an example of a little bit too much white space. There may be times when you may break long lines and write a loop statement in this way, but in this case it actually makes the code *less* readable.

Learning the Basics of Functions

Even as you learn the basics of C#, you will use *functions* in your code. A *function* is a named block of code that is intended to perform a specific task. In C#, you must write all your code inside of a function.

In the sample code you have seen so far, you have used a single function, the *Main()* function. While it is possible to write a program using only the *Main()* function, eventually you will break your code down into more manageable pieces, that is, into more than one function.

You will learn more about functions in Lesson 22, "Getting Started with C# Functions." However, to understand the sample code, you must understand the basics of functions in C#.

In C#, you must *define* a function as a member of a *class*. After you have defined a function, you may *call* the function to execute the code the function contains.

To write a function definition, you first write the function's *return type*, which is the type of value the function will pass back to the statement that called the function. For your beginning code samples, you will use a return type of *void*, which means that the function does not return a value.

After the return type, you write the name of the function. You may give a function any name you want, but usually the name will describe the function's purpose. For example, you might name a function that returns the current date as *GetCurrentDate*. Likewise, you might name a function that calculates shipping cost as *CalculateShippingCost*. When you assign meaningful names to your functions, your code becomes easier to read and understand. A function name may not contain any spaces, and it may not start with a number.

After the function name, you must give the function a *parameter list*. A parameter is a value you *pass* to the function in the statement that calls the function. For example, you might pass two parameters, a package weight and a destination, to the function that calculates shipping cost. Begin the parameter list with an open parenthesis, then write the *types* and *names* of the parameters, and end the list with a close parenthesis:

```
void MyFunction (int Var1, long Var2)
```

Even if the function does not use any parameters, you must include a parameter list by writing the open and close parentheses without any arguments:

```
void MyFunction ()
```

After the parameter list (which programmers may also refer to as the argument list), you write the *body* of the function. The body contains the statements that your program will execute when you call the function. Begin the body of a function by writing an open brace on a line by itself, then write the statements, and end the body by writing a closing brace on a line by itself:

```
void MyFunction ()
{
    Statement1;
    Statement2;
    . . .
}
```

After you declare and define a function, you may call the function simply by writing its name in your code, passing it any required parameters:

```
class MyClass
{
    static void Main()
    {
        MyFunction();
    }

    void MyFunction ()
    {
        Statement1;
        Statement2;
    }
}
```

In Lesson 22, you will learn how to write functions that use parameters and which return values to the calling statement.

Understanding the C# Class

The C# language is heavily slanted toward object-oriented programming, and the class is a key element in object-oriented programming in C#. The class groups variables and code together in a single programming type, which you then may treat as a single object.

In C#, you must declare all your variables and write all your code as **members** of a class, and so classes take on an all-important role in C# programming. To define a class, you type the keyword **class** followed by the name of the class, followed by an open brace. Then, you add the variables and code your class needs. You end the class definition with a closing brace:

```
class MyClass
{
    //  Enter variable declarations and function
    //  prototypes here.
};
```

Notice that a semicolon follows immediately after the closing brace that ends the class definition. All variable declarations and function prototypes must appear between the braces.

You will learn about declaring and using classes beginning in Lesson 7, "Getting Started with Classes."

Understanding Statements in C#

As you have learned, a *statement* is a single instruction that you write in your program. A statement is the basic building block of your programs. Without any statements, your program has no code to execute, and thus cannot perform any useful purpose.

A statement usually contains *expressions*. An expression is the basic operation unit used by your computer. You create a statement by ending an expression with a semicolon. For example, *2 + 4* is an expression. In C#, simply writing an expression with a semicolon does not make it a valid statement. The following line, while acceptable to C and C++ compilers, will cause the C# compiler to generate an error:

```
2 + 4;
```

In C#, a statement must yield a usable result. A statement in C# must modify or create an object or variable. The following are examples of valid statements in C#:

```
int x;
x = 2 + 4;
```

The first statement above creates a variable, *x*, and the second assigns a value to the variable.

Variables Store Information in Your Program

A **variable** is a memory location your program uses to store information temporarily. Your program may save the result of an expression in a variable, perform some other statements and retrieve the value from the memory location in a later statement.

When you **declare** a variable, you must give it a data type, such as **int** or **long**. The data type also may be a type that you create, such as a class object.

You will learn more about how to create and use variables in Lesson 6, "Using Variables to Store Information in C# Programs."

Using the Semicolon to Mark the End of a Statement

A C# program consists of a sequence of *statements*. You have just learned that you can break a single statement using more than one line. This means that the C# compiler must have some way to recognize when you intend to end a statement and begin the next one.

A statement ends when the compiler encounters a semicolon. This is the same as placing a period at the end of a sentence when you are writing a statement in a term paper. If there is more code on the same line, the Visual C# compiler will treat it as part of a new statement, just as you may have more than one sentence in a paragraph.

Normally, however, you will write only one statement on a line. In most cases, writing multiple statements on one line makes the code more difficult for you to read.

As you have seen, however, you can write a single statement on several lines. The compiler will put the lines together until it encounters the semicolon. To keep them *unique*, variable and function names in larger programs can become long, and writing a function call on a single line would make a very long line. In this book, you often will see a single statement written on two or three lines. For example, the following call to a function to load an image from a file is 71 characters long. You could write it on a single line as shown in the following:

```
HANDLE LoadImage(hinst, lpszName, uType, cxDesired,
cyDesired, fuLoad);
```

The function call is difficult to read and it is very hard to locate individual arguments in it. Depending upon your settings for the Visual Studio editor, the function call probably will be longer than a display line. To read the entire function call, you would have to scroll the screen horizontally, and some parts always would be out of view. If you use the fact that the compiler treats each line as part of the same statement until it encounters the semicolon, you can rewrite the line as follows:

```
hImage = LoadImage (
                hinst,       // handle to instance
                lpszName,    // name of the image
                uType,       // image type
                cxDesired,   // desired width
                cyDesired,   // desired height
                fuLoad       // load options
                );
```

White space is cheap, and spreading it around here makes the function more readable and leaves room on the right to add comments that describe each argument. You can see all the arguments on the screen at the same time. The cost is that it takes eight lines to write it this way.

Either call is OK to the compiler. Either method will cause the compiler to generate a call to the *LoadImage()* function with the arguments in the proper order.

Adding Comments to Your Code

In the last section, you saw an example of adding comments to your code when you wrote a call to the *LoadImage()* function. Adding white space to your code makes the code more readable, but it does not guarantee that you will remember *why* you wrote that particular code in that particular way.

As your programs get longer, it is important that you *document* the code by providing explanations. When you or another programmer return to the code some months in the future, you may need to jog your memory.

You can remind yourself of the program's purpose and processing by providing *comments* in your code. When the compiler encounters a comment, it ignores the text until the end of the comment. In C#, there are two methods of providing comments: the single-line comment and the multi-line comment.

The sequence of two slash marks—"//"—is the single-line comment. When the compiler encounters the two slashes, it simply ignores the rest of the line—that is, the text from the double slashes up to the newline character. There cannot be any space between the slash marks.

```
int nPos;              // Position in converted line
```

If a single-line comment extends to two or more lines, you must enter a second comment marker on each line:

```
//   This is an example of
//   single-line comments that
//   span more than one line.
```

The multi-line begins with the sequence "/*" without any space between the slash and the asterisk. When the compiler encounters the /* sequence, it ignores all text until it encounters a matching end comment marker, which is "*/":

```
/*
    This is an example of
    a multi-line comment that
    spans more than one line.
 */
```

The multi-line comment markers are handy when you must write a lengthy explanation in your code, such as to explain the purpose and result of a function:

```
/*
    Nada().
    This function does nothing. It takes no parameters
    and returns no values.
 */
void Nada ()
{
}
```

You should remember that comment markers—either the "//" or the "/*" and "*/" sequences—may not be enclosed in quotation marks, such as a text you are writing to the console screen. In the following statement, the program will write text within the comment markers—and the comment markers as well—to the console:

```
Console.Write("Hello, Visual C# /*this is not a
comment*/");
```

Newcomers to programming tend to use too many or too few comments. Some comments, such as the following, are self-evident and only confuse the text:

```
nPos = 5;            // Set nPos to 5
```

As you learn to program, it is better to err on the side of too many comments. If you are in doubt, add the comment. Comments do not cost anything other than storage space in the file, and it is far easier to remove them later than to add a missing comment.

Understanding Namespaces

In C++, you probably would not encounter *namespaces* until you reached the advanced—or at least intermediate—topics. In C#, however, it is difficult to

write any but the simplest of programs without using namespaces. C# makes extensive use of namespaces, and it is important that you understand the concept early in your studies of C#.

In this section, try to understand the *concept* of namespaces rather than concentrate on how to use them. Very shortly, you will begin learning about classes in C#, and you will have opportunities to use the namespace concept.

If you create a C# project using one of the Visual Studio wizards, the wizard will create at least one namespace, which will be the same as your project. A namespace is a convenient mechanism for holding *variables* and *functions* and other objects that have names, such as *classes* and *structures*. Within a namespace, the names of objects must be unique, but you may use the same name for an object in another namespace.

One way to look at namespaces is to compare them to the directory structure on your computer. Within a directory (the "namespace"), all the file names must be unique. Within the *C:\bin* directory, you may have only one file named *MyProg.exe*. However, in the *C:\Programs* directory, you may have another file named *MyProg.exe*. The names do not conflict with each other.

To define a namespace, use the keyword *namespace* followed by the identifier for the namespace. Enclose your object declarations in the namespace between a set of open and close braces:

```
namespace MyNamespace
{
    // Add object definitions here
}
```

To access an object in one namespace from another namespace, you *qualify* the object's name by writing the namespace name, followed by a period and then the name of the object. The following short program, *NameSpac.cs*, defines two namespaces, *First* and *Second*. Each namespace contains a *class* named *Hello*.

```csharp
/*
    Namespac.cs  — Shows how different namespaces may
            each contain classes using the same name.

            compile this program with the following
            command line:
                csc namespac.cs
 */
namespace First
{
    using System;
    class Hello
    {
        public void ShowIt()
        {
            Console.Write ("Hello, C# namespace\n");
        }
    }
}


namespace Second
{
    class Hello
    {
//
//  For all programs, Main() is the entry point. You must
//  declare a Main() function in one of your class
//  definitions.
//
        static void Main ()
        {
//
//  Create an object of the Hello class in namespace First.
//  notice that the declaration is written with the
//  namespace name, a period and then the object's name.
//
            First.Hello Show = new First.Hello();
            Show.ShowIt();
        }
    }
}
```

The *First* namespace contains a class, *Hello*, that contains the function *ShowIt()*, which actually writes to the screen. The *Second* namespace contains another class, also called *Hello*, that contains the *Main()* function, which is the entry point for the program. In *Main()*, you declare an object of the *Hello* class in the *First* namespace by prefixing the class name by *First* and then a period.

When you use the name of an object with the *namespace.object* notation, the name of the object is *fully qualified*. A fully qualified name cannot be ambiguous because a namespace may contain only a single object with a certain name.

You also may declare a default namespace with the *using* keyword. In this case, you do not need to use a fully qualified name for an object if the name does not duplicate an object in your current namespace.

In the following program, *Names.cs*, change the name of the class from *Hello* to *Print* so the namespaces do not contain duplicate names. In the *Second* namespace, you declare *First* as a default namespace with the *using First* syntax:

```
/*
    Names.cs — Shows how to access declarations in a
               different namespace with the using keyword.

           Compile this program with the following
           command line:
               csc names.cs
 */
namespace First
{
    using System;
    class Print
    {
        public void ShowIt()
        {
            Console.Write ("Hello, C# namespace\n");
        }
    }
}

namespace Second
```

```
{
    using First;
    class Hello
    {
        static void Main ()
        {
            Print hello = new Print();
            hello.ShowIt();
        }
    }
}
```

Notice that you did not have to qualify the name *Print*. Because you declared that you were *using First*, the compiler automatically searched the current namespace and the *First* namespace to find the *Print* object.

You may *nest* namespaces by defining one namespace inside of another namespace. This practice is very common in the C# system files. When you nest namespaces, you qualify the name by specifying the out namespace first, a period, and then the inner namespace, as in the following short program, *Nested.cs*:

```
/*
    Nested.cs  — Shows how different namespaces may
                each contain classes using the same name.

                compile this program with the following
                command line:
                    csc nested.cs
 */

using System;

namespace First
{
    namespace Inner
    {
        class Print
        {
            public void ShowIt()
```

```
                {
                    Console.Write ("Hello, C# namespace\n");
                }
            }
        }
}

namespace Second
{
    using First.Inner;
    class Hello
    {
        static void Main ()
        {
            Print hello = new Print();
            hello.ShowIt();
        }
    }
}
```

Notice that in the *using* statement in the *Second* namespace, you set the default namespace using the *First* namespace, then a period, and finally the *Inner* namespace.

In C# it is common to list several namespaces with the *using* keyword. In this case, the compiler will search through each of the namespaces for the names of objects you declare. In the following snippet, the compiler will search through all of the namespaces listed to find the *Console* object (which is in the *System* namespace):

```
using System;
using System.Windows.Forms;
using System.Drawing;
using System.IO;
using System.Globalization;
using System.Runtime.InteropServices;
using System.Diagnostics;

namespace MyProject
```

```
{
    public class Hello
    {
        static void Main()
        {
            Console.Print ("Hello, C#!\n");
        }
    }
}
```

The library included with C# and the Visual Studio contains more than 1,000 class definitions, and using namespaces helps you to keep your code definitions and object declarations separate from the library classes.

Making Changes to Your C# Program

As you write programs, it is important that you remember that the files containing your source code are separate from the executable files that most people consider a "program." The C# compiler creates the program file when you compile your source-code files.

When you make a change to a source-code file, the changes are not applied to the program executable file until you compile your program. You must save the changes you make, then invoke the compiler again to apply the changes to the executable file.

When you create and edit a C# project within the Visual Studio, the IDE will save any changes to your program files whenever you build the project or compile a file.

If you must save a file without running the compiler, make sure you have selected the window containing the file in the client area, then select the File menu Save option. You also can select Save All to save all the files that you have modified since your last save operation.

When you exit the Visual Studio or close a workspace, the Visual Studio will check whether you have modified any files and not yet saved them. If it finds any, it will prompt you for each file to ask whether you want the changes saved.

Recognizing Syntax Errors

Nobody's perfect. Many times through the efforts to learn programming, you will make errors. You will misspell a keyword, use an operator out of place or even capitalize a keyword (in C#, all keywords are all lower case).

A *keyword* is a word that C# has set aside to indicate a particular attribute or action. An example of keywords are *namespace* and *using*, which you encountered in the last section. *Operators* are the symbols and punctuation marks that C# uses to indicate actions to perform, such as arithmetic operations. These include such symbols as the addition and subtraction operators.

Like any language, C# has certain rules on how you use keywords and operators, and how you combine them to form complete statements. When you misuse a keyword or operator, the compiler will issue a message and refuse to compile your program. Such errors are *syntax* errors. The compiler cannot create your executable program file if your source code contains syntax errors. The compiler will identify the lines where it found syntax errors, and you may use this information to correct your source code.

In the section on Using the Output Window in Lesson 2, "Building, Running and Saving your First C# Program," you saw the effect of leaving off a closing quote in the output string. Sometimes a simple syntax error will cause more than one error message. Using the *Nested.cs* program from the last section, the following code does not have an open parenthesis in the line that writes to the console:

```
/*
    Nested.cs  — Shows how different namespaces may
            each contain classes using the same name.

            compile this program with the following
            command line:
                csc nested.cs
    */

Using System;

namespace First
```

```
{
    namespace Inner
    {
        class Print
        {
            public void ShowIt()
            {
                Console.Write "Hello, C# namespace\n");
            }
        }
    }
}

namespace Second
{
    using First.Inner;
    class Hello
    {
        static void Main ()
        {
            Print hello = new Print();
            hello.ShowIt();
        }
    }
}
```

Although it is a single error, the compiler issues three different messages:

```
Nested.cs(20,19): error CS1002: ; expected
Nested.cs(20,42): error CS1002: ; expected
Nested.cs(20,42): error CS1525: Invalid expression term ')'
```

The missing parenthesis causes the compiler to lose track on the statement, and so it issues errors until it can recover. When you receive multiple error messages after compiling your program, always begin work on the first error listed.

WHAT YOU MUST KNOW

In this lesson, you have learned some of the basics of programming in C#. This is only a start, and in the following lessons you will learn more about how to declare and use objects and how to use variables, functions and operators. You also learned in this lesson how to use white space to make your code more readable, and how to add comments to your source-code files. You also learned about the namespaces, which is an important concept in C# programming. You learned how to save changes to your source code and how to recognize syntax errors issued by the compiler. In Lesson 6, "Using Variables to Store Information in C# Programs," you will learn how to declare and use variables in a C# program. Before you continue with Lesson 6, however, make sure you have learned the following key concepts:

- ✗ White space is any character that causes cursor movement on your screen but does not display any character, such as the space or tab characters.

- ✗ Tokens are the C# keywords and operators and any other object that your program defines. C# uses white space to separate tokens.

- ✗ A statement is a single instruction that you write in your program. It may span several lines. In C#, a statement always ends with a semicolon.

- ✗ The double slash sequence (//) indicates a single-line comment to the compiler. When the compiler encounters a double slash, it ignores all the remaining text on the line.

- ✗ A multi-line comment begins with the "/*" sequence. When the compiler encounters this combination of characters, it ignores all of the text until it finds the end of comment sequence, "*/". Comments between the begin and end sequences may span several lines.

- ✗ Namespaces are convenient groupings to hold the names of objects in your source-code program. Namespaces let you group object definitions and declarations without having to worry about name uniqueness. C# makes extensive use of namespaces.

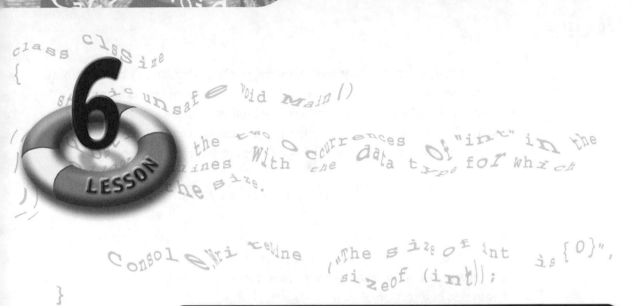

USING VARIABLES TO STORE INFORMATION IN C# PROGRAMS

more likely than not, any program you write, even simple programs, will need to store information in a form your program can use in statements and pass to other functions. When your program performs an operation—say, you multiply 42 by 24—the result is temporary. Your computer's central processing unit stores the value in a temporary location, which it overwrites in the next operation. To save the result for later use, your program must use *variables*, named memory locations where your program saves and retrieves values. In this lesson, you will learn how to declare and use variables in C#. By the time you finish this lesson, you will understand the following key concepts:

- An *identifier* is a name you use in your C# program to represent a variable, a function or a definition. C# uses identifiers to keep track of your variables and functions when it compiles your program.
- A variable represents a memory location where your program stores a value.
- The data type you assign to a variable determines what kind of value the variable will hold and what operations you may perform on the value.
- Before you can use a variable in a C# program, you must *declare* and *initialize* the variable. The declaration tells the compiler how much memory to set aside to hold the value. You initialize a variable by giving it a starting value.
- C# divides variables into *reference*-types and *value*-types. You create reference-types using the *new* operator.

- An *object* is a variable that may contain *methods* (functions) and attributes (properties and variables). C# treats all values, even numbers, as objects.
- When you assign a value to a variable or perform an arithmetic operation, you may use the *checked* keyword, which causes your program to generate an error when you try to store a value that is too large or too small for the memory location.

Understanding Identifiers

One of the advantages of a programming language is that you may refer to memory locations using names. To a computer, a memory location is just a number that marks the *address* in memory where you store a value, or which marks the beginning of a section of code.

A memory address is not unlike the address of your house. When someone writes you a letter, it is not enough for the letter writer simply to put your name on the envelope. To direct the letter through the postal system, the envelope must have your address, which directs the letter to a particular city, then to a particular street and finally to a particular house on that street. In the same way, a memory address directs the computer to a location in its memory where you want to store, or already have stored, a value.

If you had to keep track of memory addresses yourself, you quickly would become bogged down in a maze of numbers. Instead, by using a programming language, you may a assign a name (create an *identifier*) for a memory address. The compiler, in turn, will match the identifier with the memory location when you compile your program.

For example, when you write your source code, you may use the name *Salary* to represent the amount an employee is paid per hour. The compiler then sets aside enough memory to hold that value. Later, when you use the word *Salary*, your program will look in that memory location to store or retrieve the employee's salary.

Identifiers may be virtually any word that you want to assign to the memory location, but you should remember that C# is a *case-sensitive* language. If you give a variable the name *Salary*, you cannot later refer to it as *salary* with a lower-case "s". The two words would refer to different variables.

An identifier may not begin with a number or any of the special characters reserved by C# as operators. It must begin with a letter—either upper- or lower-case—or an underscore character ("_"). Also, the name may not contain any of the C# operators. For example, if you try to name a variable *Mine+Yours*, the compiler would try to find the value of the variable *Yours* and add its value to the value of the variable *Mine*.

Why C# Uses Variable Types

The kind of value you store in a variable determines the variable's *data type*. The data type then determines the range of values you may store in a variable of a particular data type. For example, a *char* type requires two bytes to hold a single character. A *long*, on the other hand, requires eight bytes to store a single value. If you declare a variable of type *char* and later try to store a *long* value in that variable, you would lose a part of the *long* value.

For this reason, C# is a *typed* language. In other places, you may see C# described as "strongly typed," but compared with other languages it is difficult to make an argument for that description. In C#, you cannot store a value of one type in a variable that you have declared as another data type, but there are plenty of exceptions. The C# compiler enforces these rules, and will issue an error when you attempt such an operation.

Understanding Fundamental C# Types

Because it is a typed language, C# defines a basic set of data types that you may use. These are the *fundamental* data types. The fundamental types provide storage for numbers of different sizes. In addition to the fundamental types, you also may declare your own data types. The types you create are *derived* data types. You create a derived data type when you define a class or structure in your C# program.

Table 6.1 summarizes the fundamental data types in C# and gives their sizes in bytes. Remember that these are the sizes on the Windows operating system; if C# is implemented on other operating systems such as UNIX or the Mac OS, these sizes might be different.

Data type	Size (bytes)	Description
byte	1	Unsigned byte
sbyte	1	Signed byte
bool	1	Boolean value (true or false)
char	2	Unicode character
string	varies	Unicode string
short	2	Signed short integer
ushort	2	Unsigned short integer
int	4	Signed integer
uint	4	Unsigned integer
float	4	Floating point number (contains a decimal point)
long	8	Signed large integer
ulong	8	Unsigned large integer
double	8	Double-precision floating point
decimal	16	Fixed precision floating-point value typically used for calculations involving money

Table 6.1 The pre-defined data types used by C# and the number of bytes required to store the values.

Because C# is a new language and is still evolving, you may see different values elsewhere. For example, I have seen the size of the *decimal* type variously as four, eight and 12 bytes. However, if you use the *sizeof* operator to obtain the size, the current version of C# reports 16 bytes.

If you need to know the size of a data type, you can use the following short program, *Size.cp*, that uses the C# *sizeof* operator to display the number of bytes of storage a specific data type (such as *int*) provides. To compile code from the command line that you mark as unsafe, you must include the */unsafe* switch:

```
/*
    size.cs. Use this program to determine the size of a data
         type in C#.

    Compile this program using the following command line:
         csc /unsafe size.cs
*/
using System;

class clsSize
{
```

```
    static unsafe void Main()
    {
//  Substitute the two occurrences of "int" in the
//  following lines with the data type for which
//  you want the size.
//

        Console.WriteLine ("The size of int is {0}",
                            sizeof (int));

    }
}
```

Although the C# documentation lists *string* as a basic data type, you will recall from Lesson 4, "Using C# From the Command Line," that *string* actually is an alias for the *System.String* class. C# will not let you use the *sizeof* operator on a class, so if you substitute *string* for *int* in the preceding program, the compiler will issue an error.

Note: *The program size.cs uses the unsafe attribute within the function declaration for Main. As it turns out, to use the sizeof operator, you must specify the code as "unsafe." You will learn more about the unsafe keyword in Lesson 9, "Getting Started with C# Operators."*

Declaring Variables Within a C# Program

Before you can use a variable to store a value in C#, you first must *declare* the variable. The declaration tells the C# compiler the variable's data type, where to create the variable (on the stack or on the heap), and how much memory to set aside to hold a value.

Variables in C# may be broken into two broad classifications: *value* and *reference*-types. The data type and the method you use to declare and create a variable will determine whether it is a value-type or a reference-type. You create a value-type variable on the *stack*, and a reference-type variable on the program's *heap*.

To declare a value-type variable, simply write the data type for the variable followed by the name—or identifier—you want to use for the variable. The following statement, for example, declares an integer variable:

```
int iVar;        //  Declares a value-type variable
```

When you declare a value-type variable, the program sets aside enough space on the program's stack to hold the variable's value. The stack is a section of memory that your program maintains to hold temporary information. When the function in which you create a value-type variable ends, the stack is adjusted so that the variable no longer exists.

To declare a reference-type variable, you must use the *new* operator. To begin, write the data type and identifier for the variable just as you would a value-type. Follow the identifier with an equals sign, the *new* operator, the data type and finally a set of parentheses:

```
clsClassName rVar = new clsClassName();
```

The syntax might look a little strange if you are used to declaring variables in C or C++, but in essence you are creating a "managed" object on the heap. After you create a managed object, the Common Language Runtime (CLR) module, which Windows loads when you run a C# program, begins managing the variable. When the Common Language Runtime determines that you no longer need the information in the variable, it destroys the variable.

Of course, there's little advantage to creating a managed object for a simple data type, but when you study classes in the next lesson, you will see that it has its benefits. In fact, the only way you can declare an object of a class is to use the reference method.

Indentifiers May Not Duplicate C# Keywords

Keywords are identifiers that have special meaning to the C# compiler. When you compile a program, the compiler will perform a predetermined action when it encounters a keyword. You cannot use any of these predefined keywords as identifiers for variables or objects in your program. Table 6.2 lists the C# keywords. As you can see, the C# keywords are all lowercase.

abstract	as	base	bool	break
byte	enum	case	catch	char
checked	class	const	continue	decimal
default	delegate	do	double	else
event	explicit	extern	false	finally
fixed	float	for	foreach	goto
if	implicit	in	int	interface
internal	is	lock	long	namespace
stackalloc	new	null	object	operator
out	override	params	private	protected
public	readonly	ref	return	sbyte
sealed	sizeof	static	string	struct
switch	this	throw	true	try
typeof	uint	ulong	unchecked	unsafe
ushort	using	virtual	short	void
while				

Table 6.2. C# reserves these keywords for internal use. You may not use them as identifiers.

*If you really must use a reserved keyword as an identifier—perhaps you are translating a program from another language to C#—you may prefix the name with an "at" sign. C# then will recognize the name as a unique identifier and not as a keyword. For example, **virtual** is a C# keyword but **@virtual** is not.*

Use Meaningful Variable Names

When you declare variables within your C# programs, choose names for your variables that meaningfully describe the information the variable contains. Although C# will consider the variable names such as x, y, and z valid, another programmer who is reading your code would better understand names such as **EmployeeName**, **OfficeNumber**, **EmailAddress**. Further, when you declare variables, place a comment to the right of the declaration that explains the variable's purpose, as shown here:

```
int NumberOfEmployees;    // The total number of employees
                          // working at the plant

int AverageEmployeeAge;   // The average age of the plant's
                          // employees

float AverageSalary;      // The average annual salary of
                          // the plant's employees
```

By assigning meaningful variable names and by using comments to describe each variable's purpose, you will make your program easier for other programmers to read and understand.

Assigning Values to Variables

In C#, you cannot use a variable until you assign a value to the variable. If you try to use a variable before you give it a value, the C# compiler will issue an error and will not compile your program.

After you have declared your variable, you can assign it a value by using the assignment operator, the equals sign. In this case you are *initializing* the variable, or giving it a starting value:

```
int x;
x = 42;
```

The statement in which you declare the type of the variable is a *declaration statement*. The second statement is an *assignment statement*.

You also may assign a value to the variable when you declare it, in which case the statement is a *declaration assignment*:

```
x = 42;
```

Displaying a Variable's Value

Often in programming you must display the value of a variable. You may want to show the user the result of adding a column of numbers, or you may want to display the text you have stored in a string in your program.

Earlier in this lesson, in the section on "Understanding Fundamental C# Types," you saw an example of writing an unknown value to the screen:

```
Console.WriteLine ("The size of int is {0}",
                   sizeof (int));
```

The quoted text within the parentheses is the *format string*. The *Write()* and *WriteLine()* functions examine this string to determine what to write to the screen. When the function encounters a number within braces, such as the "{0}", the function looks for a matching parameter following the format string. A {0} (remember that count in C# begins at 0) indicates that the first following parameter should be written to the screen. A {1} would instruct the function to write the second parameter to the screen. In each case, the function replaces the sequence of opening brace, number, and closing brace with the actual parameter.

You must be careful that there is a matching parameter for each number. If there is not a matching parameter, your program will throw an exception. If your program throws an exception and you do not handle the exception, your program will end abruptly.

In the *Size.cs* program in this lesson, try changing the "{0}" to "{1}" without changing the number of parameters. Compile and run the program from the command line. The program will throw an exception, but because you have installed Visual Studio on your computer, the program will display a dialog box as shown in Figure 6.1, and ask whether you want to debug the program at the point where the exception occurred. This feature of Visual Studio is "Just-In-Time Debugging." On a computer without Visual Studio, the program would end with an error message.

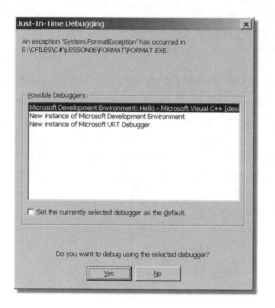

Figure 6.1. The Just-In-Time Debugging window appears when a program throws an exception on a computer where you have installed Visual Studio.

Click on the No button to stop the debugging process. Be sure to change the text back to "{0}" and to save your change.

The sequence between the braces actually may be a fairly involved expression. In addition to the argument number, you also may specify *field width* and formatting characteristics as shown in Figure 6.2.

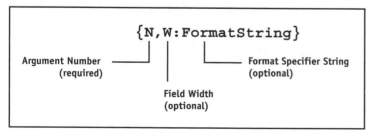

Figure 6.2. The general form of the C# format gives you the capability to set the field width and to specify how the output should appear.

The **N** in Figure 6.2 is the 0-based argument number. The **W** is a number specifying the minimum number of spaces in which you want the output to appear. If **W** is positive, the argument will be printed to the right side of the field. If **W** is negative, the argument will display on the left side of the field. If the argument is smaller than the field width, the blank parts of the field will contain spaces. If the field is too small to contain the argument's text, the entire argument will display regardless of the field width.

If you do not care about the actual width of the field, you may omit the width value entirely or specify 0. The following program, *Width.cs*, demonstrates how the width value works:

```
/*
    Width.cs. Uses the width specifier to write to the console.

    Compile this program using the following command line:
            csc Width.cs
 */
using System;

class clsWidth
{
    static void Main()
    {
        Console.WriteLine("Width is {0,6} characters", 6);
        Console.WriteLine("Width is {0,-6} characters", 6);
    }
}
```

When you run this program, you will see the following output. The positive width leaves five spaces before the argument, and the negative width forces the spaces to occur after the argument:

```
Width is       6 characters
Width is 6       characters
```

After the width value, you may write a colon (":") followed by an optional flag in the format string. You may follow the flag character with an optional *precision* value, which will vary according to the flag. Table 6.3 summarizes the formatting characters you may use in the format strings and how the flags operate.

Character	Description	Precision
C or c	Local currency format	Normally two digits past the decimal point.
D or d	Integer	If you supply a precision, the field will be padded with leading 0's.
E or e	Scientific	The precision sets the number of decimal places (the default is 6). There is always at least one digit after the decimal point.
F or f	Fixed point	The precision specifies the number of decimal places. May be 0.
G or g	General	Use E or F formatting, whichever is more compact.
N or n	Number	Formats the number with embedded commas if the value exceeds 999. A number following this flag specifies the number of decimal places to print.
X or x	Hexadecimal	Formats a hexadecimal (base 16) number using the precision specifier.

Table 6.3 The formatting characters used to write arguments.

The formatting flags, combined with the width and precision numbers, give the C# formatting nearly the same power as the C and C++ formatting sequences. To see how these formatting characters work, enter and run the following program, *FormTest.cs*:

```
/*
    FormTest.cs. Compares the different formatting flags
                 for writing text to the console screen.

    Compile this program using the following command line:
             csc FormTest.cs
*/
using System;

class clsFormat
{
    static void Main()
    {
```

```
Console.WriteLine ("The cost is {0:C}", 2);
Console.WriteLine ("The cost is {0:C}", 2.36);
Console.WriteLine ("The value is {0,0:D5}", 1024);
Console.WriteLine ("The value is {0,0:N5}",
                      1024742);

Console.WriteLine ("The value is {0,0:N}",
                      1024742);
Console.WriteLine ("The value is {0,0:N5}",
                      1024742);
Console.WriteLine ("The value is {0,0:X}", 1024742);
Console.WriteLine ("Pi is {0,0:F3}", 3.14159);
Console.WriteLine ("Pi is {0,-8:F5}", 3.14159);
Console.WriteLine ("Pi is {0,-8:E2}", 3.14159);
Console.WriteLine ("Pi is {0,-8:E}", 3.14159);
    }
}
```

Notice that on the first write to the control, you pass the *Console.WriteLine* function the format string and the number 2. When the line prints, the function adds the dollar sign and two decimal points to the output. The following is the output from *FormTest.cs*:

```
The cost is $2.00
The cost is $2.36
The value is 01024
The value is 1,024,742.00000
The value is 1,024,742.00
The value is 1,024,742.00000
The value is FA2E6
Pi is 3.142
Pi is 3.14159
Pi is 3.14E+000
Pi is 3.141590E+000
```

Instead of the width and format specifiers, you may create a *picture* of the output using a sequence of # symbols or 0's and punctuation marks as in the following program, *PicForm.cs*:

```
/*
    PicForm.cs.   Compares the different formatting flags
                  for writing text to the console screen.

    Compile this program using the following command line:
          csc PicForm.cs
*/
using System;

class clsPicForm
{
    static void Main()
    {
        Console.WriteLine ("The cost is {0,0:$###.##}", 2);
        Console.WriteLine ("The cost is {0,0:$###.##}",
                          2.36);
        Console.WriteLine ("The value is {0,###,###}",1024);
        Console.WriteLine ("The value is {0,#,###.#####}",
                          1024.742);

        Console.WriteLine ("The cost is {0,0:$000.00}", 2);
        Console.WriteLine ("The cost is {0,0:$000.00}",
                          2.36);
        Console.WriteLine ("The value is {0,0,000,000}",
                          1024);
        Console.WriteLine ("The value is {0,0,0,000.000}",
                          1024.742);
    }
}
```

There are many, many more combinations of the formatting flags and values. Take some time to experiment with them. When you finish, here is some good news: everything you learned about formatting for console output may be used to format strings in your program. You may want to do this from time to time to pass formatted text to a function, such as to display a message box.

To format a string, simply replace the *Console.Write* with *String.Format* and assign the formatted string to a *string* variable:

```
string str;
str = String.Format ("The value is {0,0:D5}", 1024);
MessageBox.Show (str);
```

Or, in the case of an assignment declaration, you may format the string when you declare the variable:

```
string str = String.Format ("The value " +
                            "is {0,0:D5}", 1024);
MessageBox.Show (str);
```

Using Just-In-Time Debugging

When you debug a program in the Visual Studio, you are running the program in a protected environment. The debugger initializes some variables and conditions that might not be present when you run your program outside of the debugger.

When you are developing an application, you must test-run the program periodically outside of the debugger to test the program under real conditions. When you compile your program, the C# compiler inserts information into the executable program that tells Windows what to do when the program encounters an error that prevents the program from continuing.

On your test system—the one on which you installed the Visual Studio—Windows will ask if you want to debug the program using Visual Studio. When Windows displays the Just-In-Time Debugging dialog box, if you click on the Yes button, Windows will start the Visual Studio and load your program into the debugger.

To take advantage of the capabilities of Just-In-Time Debugging when you compile a program from the command line, you need to add debug information to your executable file. You do this by specifying /debug on the command line when you compile your program.

*To compile the **FormTest.cs** program from the command line and include debugging, you would use the following command line:*

```
C:> csc /debug FormTest.cs  <Enter>
```

If your program "crashes" and you start the debugger, the Visual Studio will take you directly to the line that contains the problem.

Omitting the "0:" sequence from a format string is almost certain to cause the program to throw an exception. Try omitting just one "0:" from just one **Console.WriteLine()** *statement, compile the program with the /debug flag and run the program. Follow the Windows prompts and the Visual Studio will take you directly to the line containing the error.*

Assigning a Value of One Type to a Variable of Another

Within your programs, you may experience errors when you try to assign a value of one type (such as *int*) to a variable of a different type (such as *char*). When you try to perform such an assignment, the C# compiler will generate an error message. The compiler generates the error because your assignment might cause part of a value to be lost because the smaller variable type cannot store the large value.

For example, try the following simple program, *Assign.cs*, to get an idea of the error messages the compiler might issue.

```
using System;
class clsAssign
{
    static void Main ()
    {
        char A_char = 'A';
        long A_Long = 42;
        A_Char = A_Long;
    }
}
```

In this case, the code creates two variables, one of type *char* and one of type *long*. The code uses the assignment operator to assign values to each variable. Then, the code tries to assign the value of the *long* variable to the *char* variable. If you try to compile this program, the compiler will tell you that it cannot *implicitly* convert the *long* value to a *char* value:

```
Assign.cs(9,17): error CS0029: Cannot implicitly convert type
'long' to 'char'
```

In cases where your code must assign a value of one type to a variable of another, you can force the compiler to perform a conversion by *casting* the larger value to the smaller value. To cast one data type to another, simply write the new data type inside parentheses before the value or variable:

```
A_Char = (char) A_Long;
```

Casting is an *explicit* conversion of one data type to another. You should be aware that in casting a larger value to a smaller value, you risk losing part of the value. Your program will discard the bytes that it cannot use when it makes the conversion and assigns the value to the smaller data type.

However, C# will *implicitly* convert some values when that conversion would not cause you to lose any information. In the above example, you could assign the *char* value to the *long* value without causing the compiler to generate an error:

```
A_Long = A_Char;
```

The size of the *long* variable is larger than the size of the *char* variable, so no information is lost in this assignment. In a statement such as the preceding, if the value on the right side of the equals sign is unsigned and the value on the left is a signed type, the new value will be positive. If, however, both are signed types, the result is the sign of the value on the right side of the equals sign.

Using Read-Only Variables

There are times when you must set a value in your program that you will not want to change—a *constant* value. C# provides two methods to make the value in a variable *read-only*.

First, you may use the *const* keyword when you declare a variable. In this case, you must assign a value when you declare the variable:

```
const double Pi = 3.14159;
```

If you later try to change the value of a *const* variable, the compiler will issue an error and will not compile your program.

The problem with a *const* is that you must assign the value at the time you compile your program. There are times when you will not know the value and must calculate it when your program runs. When you must do this, you may use the *readonly* keyword when you declare your variable. The following program, ReadOnly.cs, uses the *readonly* keyword to make the variable *area* readonly. The program calculates the area of a circle based on the radius value you specify in the command line when you run the program:

```
/*
    ReadOnly.cs. Demonstrates the use of a readonly
                 variable.

    Compile this program using the following command line:
            csc ReadOnly.cs
*/
using System;

class clsMath
{
    readonly double area;
    const double pi = 3.14159;
```

```
clsMath(double arg)
{
    area = arg * arg * pi;
}

static void Main (string[] arg)
{
    if (arg.Length != 0)
    {
        clsMath math = new clsMath(Convert.ToDouble(arg[0]));
        Console.WriteLine ("The area is {0}",
                                math.area);
    }
}
}
```

When you use the *readonly* keyword to declare a variable, you may set the value only once. After the *readonly* variable's value has been set, the compiler will not let you change it in a later statement. And there is a catch. You must assign the variable a value when you declare the variable, or in a *constructor* function. A constructor is a special function that executes whenever you declare a class object and always has the same name as the class.

If you assign the value when you declare a *readonly* variable, the variable essentially is the same as a *const* variable.

You will begin to learn about classes in Lesson 7, "Getting Started with Classes," and how to set up and use constructors in Lesson 10, "Understanding Constructors and Destructors."

C# Manages Some Data Types

*C# will not let you use the sizeof operator on a class definition because in C# a variable of a class type is a **managed** variable. Managed variables are those that the Common Language Runtime allocates and releases automatically.*

*You may create a managed variable only by using the **new** operator. This causes the program to place the variable's memory in the **heap** (sometimes called the **free*** *store). The heap is a block of memory the operating system reserves for your pro- gram to create variables and objects for which the size cannot be determined until the program is actually running.*

*Suppose, for example, you have defined a class named **clsStuff**. You must create an object of the **clsStuff** class using the **new** operator:*

```
clsStuff MyObject = new clsStuff();
```

*You will learn the meaning of the declaration syntax in Lesson 7, "Getting Started with Classes." For now, you should understand you must use the **new** operator to create class objects.*

*Unlike C and C++, when you create an object on the heap using the **new** operator, you do not have to worry about deleting the object when you are finished with it. The management properties of C# will delete the object when your program no longer needs the object.*

Using Checked and Unchecked Variables

The range of values that you may store in a C# variable is very large. The range depends upon the data type and whether you declare a variable as *unsigned*. A variable of type *int*, for example, may store values from 2,147,483,647 to –2,147,483,648. Although the range is very large, the upper limit still is less than the size, in bytes, of most hard drives available on a typical computer.

It is quite possible, then, that your calculations could make your program try to store a number that is too large (an *overflow* condition) or too small in the case of *double* or *float* values (an *underflow* condition). How your program responds to these conditions depends upon several things, including compiler flags that you may set when you build your program. The default is for your program not to check for overflow and underflow conditions. You may change the default by adding the "/checked" flag to your command line:

```
C:> csc /checked ReadOnly.cs   <Enter>
```

Regardless of the default setting, you can make your program behave in one particular manner by using the *checked* and *unchecked* keywords. You may specify one of these keywords for a particular statement, or for a block of statements. When your program executes the following statements, it will throw an exception because the result of the expression will be larger than the maximum value you may store in an *int* variable:

```
int DiskSize = 2147483647;        // 2 GB disk
int BiggerDisk = checked (4 * DiskSize);
```

To prevent the exception, you would use the *unchecked* keyword:

```
int DiskSize = 2147483647;        // 2 GB disk
int BiggerDisk = unchecked (4 * DiskSize);
```

When you use the *unchecked* keyword, the result of an operation almost certainly will be incorrect if an overflow or underflow condition results, and you would need to check the result in your code.

You also may apply *checked* and *unchecked* to a block of statements:

```
checked
{
    int BigDisk = 2147483647;        // 2 GB disk
    int BiggerDisk = checked (4 * DiskSize);
}
```

The *checked* keyword will apply to all of the statements between the opening and closing braces.

Boxing and Unboxing Variables

As you will learn, C# treats all variables as *objects*. Within your C# programs, you may perform class-like operations on value-type variables, and on numbers as well.

Somewhere deep within C# is a class definition named *object*. Ultimately, C# derives all variables and all values from the *object* class. In cases where you must make a value-type variable act like an object, your programs may call *object*-class functions to perform those operations. When your programs use the *object*-class functions, programmers will say that you are "putting the value into an object *box*." The process of using the *object*-class functions with a value is called *boxing*. For example, say you need to write the value 4096 to your screen. You can use the *Console.Write()* function, which requires a string to display, using a statement such as the following:

```
Console.Write ("{0}", 4096);
```

In this case, the *Console.Write()* function will convert the value, based on the format specifier into a character string for display. The *object* class, however, contains its own conversion function, *ToString()*, so you could write something like the following:

```
Console.Write (4096.ToString());
```

Just as you can box a value-type variable, there are times that you want a reference-type variable to act like a value-type. The reverse of boxing is *unboxing*:

```
int value;                  // Create the variable
value = 4096;               // Give the variable a value
object someObj = value;     // Box it
[some statements that use the object]
int y = (int) someObj;      // Unbox it.
```

Unboxing is inherently more dangerous than boxing. When you unbox a reference variable, the management code first checks to see whether the variable contains the type you want to take out of the box. If it does not, the program will "throw an exception." If you do not handle the exception (you will learn how in Lesson 15, "Handling Exceptions"), your program will display some dire error messages and end abruptly.

WHAT YOU MUST KNOW

In this lesson, you have learned how to declare and use variables of different types in C#. You have learned how C# treats all values, even numerical constants, as objects. When you apply an object method to a variable, you are *boxing* the variable. When you make an object appear as a value, you are *unboxing* the object. You also learned about the *System.Console()* function and how to format and display text on your screen, and to apply the same formatting techniques to building a string in your program. You learned about read-only variables and how to check for overflow and underflow conditions when assigning a value to a variable. In Lesson 7, you will begin studying C# classes and learn how to use them in your program. Before you continue with Lesson 7, however, make sure you have learned the following key concepts:

- A variable is a named memory location in which your program may store information.

- Before you may use a variable, you must declare the variable using the type of value—such as *int*, *double* or *float*—that you want to store in the memory location. This tells the C# compiler how much memory to reserve for the value.

- You must give a variable a starting value before you may use the variable in a C# expression.

● Variables in C# are *value*-types or *reference*-types. Your program creates value-types on the stack and reference-types in the heap or free store.

● After you set the value of a *readonly* variable, your program may access the value but it does not change the value.

● When you store a value that is too large or too small for a variable, your program may generate an error. You can control this behavior using the *checked* and *unchecked* keywords.

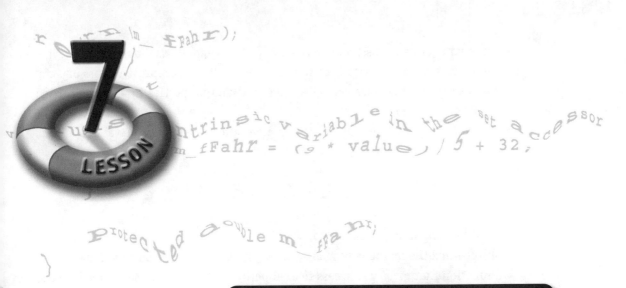

GETTING STARTED WITH CLASSES

C# is an *object-oriented* language, and the *class* is the primary mechanism that you will use to create objects in a C# program. In fact, the class is a key element to C# programming. Without at least one class, you cannot have a C# program. Classes usually contain variables and properties in which your program may store and retrieve information. A class also may contain functions, or *methods,* that you use to set, retrieve and modify the information contained in a class. In C#, all variables and all functions must be *members* of a class; you cannot declare a variable or define a function outside of a class definition. In previous lessons, you have used classes in your beginning programs without knowing how the class mechanism works. In this lesson, you will begin to learn about the C# class and how it works. You will learn how to define your own classes and to add properties and methods. By the time you finish this lesson, you will understand the following key concepts:

- In C#, the *class* is the basic unit of object-oriented programming.
- A class definition provides the compiler with a template. The compiler uses that template when your program creates an object using the class definition.
- To define a class, you specify the name of the class using the *class* keyword. You then declare the variables (properties) and functions (methods) that will be members of the class.
- All variables that you declare in a C# program, and all functions that you define, must be members of a class.

- When you define a class, you create a new data type—a *user-defined* data type. You may use this new data type to declare variables that represent class objects.
- You may declare member variables, functions and properties in a class by using the Class View tool window.
- The *static* keyword on class members gives those members special properties, and special restrictions.

Introducing the C# Class

In object-oriented programming, you focus on the properties of a system—or object—and the methods to modify or manipulate those properties. A method is a function that gives you access to a property so that you can set or retrieve the property's value, or perform an operation on the value.

In C#, the class is the primary mechanism—the building block—for object-oriented programming. In fact, C# is very object-oriented, and your program must have at least one class or you cannot write any code.

A class lets you bind data and functionality together into a single object. You can set the access level for the various functions and variables that are members of the class, thus making some of them available to the entire program and some for use internally by the class.

A well-designed class is *reusable*. In a single program, you may declare multiple instances of the class, and you may copy a class and use it in another program. In C++, it is common for programmers to group related classes and compile them into a library. In this way, other programs can reuse the classes without having to compile the class code again. Although C# is a new language and nothing really is common practice yet, the properties of the .NET runtime make this possibility highly likely.

In C#, classes are the containers for all code and variables. You cannot declare variables or define functions outside of a class definition. In C#, all variables and all functions are *members* of some class within your program. A class may contain special functions known as *constructors*. A constructor runs automatically when you create a class object. Using a constructor function, a class can initialize itself (by assigning values to its member variables). Another special function, a *destructor* (also called a *finalizer* in C#), runs automatically when the runtime module destroys a class object.

A class is a *reference* object, and the only way you may create an instance of a class object is in the heap by using the *new* operator. After you create a class object, the Common Language Runtime manages the memory occupied by the class object, and frees the memory when the class object is no longer needed. This differs from C++, in which it is the programmer's responsibility to free objects on the heap.

Later, you will learn about another object, the *structure*, which is very similar to a class, but is a *value* object. When you declare an instance of a value object, C# will create it on the stack like an ordinary variable rather than on the heap. Whether you use a class or a structure will depend upon the needs of your program.

Defining Classes

In C#, a class definition also implements the class. You define and implement the class in the same file. In this way, C# is more like Java than C++. It is common in C++ to define a class in one file, and then to implement the class code in another file using the #include preprocessor directive to read the definition file. C#, however, does not have or need a #include directive.

A class definition begins with the keyword *class*. You may use a *modifier keyword* before the *class* keyword, and you will learn about these shortly. After the *class* keyword, you enter an open brace symbol, **{**. You should write this symbol on a line by itself so that it stands out in your code. You then begin adding the variables and functions (the *properties* and *methods*) that your class will use.

When you have finished entering the properties and methods, type a closing brace, **}**, on a line by itself. If you have studied C++, you probably have learned that you must type a semicolon following the closing brace. You may read elsewhere that including a semicolon here will cause a compiler error, but that specification apparently has changed since the original introduction of C#. The compiler does not seem to care, and the C# language specification released by Microsoft lists the semicolon as "optional." If you want to maintain some con-

sistency between C++ and C#, you may write a semicolon without causing a compiler error despite what you read elsewhere.

The following is an example of a bare class definition:

```
class clsMyClass
{
    //   Variable declarations
    //   Function definitions
}   // or };
```

Notice that the class name, or identifier, begins with the letters "cls." The compiler does not require this sequence, and so far no clear "standard conventions" have evolved for writing class names in C#. However, in Visual Basic, it is common practice to precede a class name with the letters "cls." Some C# examples I have seen use a capital C in front of the class name, which is the convention in Visual C++. Neither convention is incorrect. Whichever you use, try to be consistent so that you may recognize a class name easily in your code.

You may have as many classes as your program needs, but each class definition must have a unique name within the namespace. You may duplicate class names in different namespaces. The following snippet declares two classes by the same name but in different namespaces:

```
namespace nsFirst
{
    class clsMyClass
    {
        //   variable and code here
    }
}
namespace nsSecond
{
    class clsMyClass
    {
        //   variable and code here
    }
}
```

You encountered namespaces in Lesson 5, "Understanding a Few Key C# Basics," and you will learn more about using namespaces in Lesson 13, "Using References and Assemblies."

Now that you have defined a skeleton class, you are ready to add member variables and functions.

Adding Members to a Class

All the functions and variables that your program will use must be members of a class. To use a variable, you must declare it in the class definition. Within a class definition, the C# compiler does not care in what order you declare and use variables. You may use a variable in a member function so long as you declare it somewhere in the class, even at a point after where you first use the variable.

To declare a variable in a class, you first type the *access* keyword for the variable. C# defines five access levels for class members. You will learn more about access keywords in Lesson 11, "Understanding Class Scope and Access Control." For now, however, declare your variables using the *public* access level. Next you type the data type of the variable—whether it is an *int*, a *long*, a *double*, a *float*, and so on. Then, you write the name of the variable followed by a semicolon, creating a *declaration statement*. The following lines declare variables of different types in *clsMyClass*:

```
class clsMyClass
{
    public int iVar;
    public long lVar;
    public double fVar;
}
```

A variable name must be unique *within the class*. However, you may have variables of the same name in different classes. The following snippet defines two classes, *clsClassOne* and *clsClassTwo*, and declares a variable of the same name in each class. The two names do not interfere. Later in this lesson you will learn how to access each variable uniquely:

```
class clsClassOne
{
    public int iVar;
}
class clsClassTwo
{
    public int iVar;
}
```

Before you may use a variable in C#, you must give the variable an initial value. You can initialize the class variable when you declare the variable, or you can initialize the class variable within a class method (function). You must, however, assign the class variable a value before you use the variable. The following code, for example, will cause the compiler to issue an error because the variable *iVar* does not have an initial value in *MultiplyVar()*:

```
class clsVarTest
{
    public int iVar;
    int AddVar(int iParm)
    {
        iVar = 2;      // OK. iVar now has a value
        int result = iVar + iParm;
        return (result);
    }
    int MultiplyVar(int iParm)
    {
        // Error. iVar does not have an initial value in
        // the following statement
        int result = iVar * iParm;
        return (result);
    }
}
```

To make the above code work, you would have to give *iVar* an initial value in *FunctionTwo()*.

As discussed, you can assign a variable an initial value when you declare the value. A statement that both declares and initializes a variable is a *declaration assignment statement*. The following code will compile because *iVar* has an initial value in both functions:

```
class clsVarTest
{
    public int iVar = 2;
    public int AddVar(int iParm)
    {
        int result = iVar + iParm;
        return (result);
    }
    public int MultiplyVar(int iParm)
    {
        int result = iVar * iParm;
        return (result);
    }
}
```

Giving Class Members an Access Level

In most of the examples so far, your variables and functions have worked without access keywords because they were all declared and defined within a single class. After this lesson, however, your programs are going to be more complex and generally will involve using more than one class. The access keywords will become more important when you attempt to use variables and functions in other classes.

*C# defines two access levels for classes, **public** and **internal.** The default access level is **public.***

*The **public** keyword gives virtually unlimited access to a variable or function. Any function in any class may access the variable or function.*

*Setting an access level of **internal** is the same as **public,** except that it limits access by functions in classes within the current project only. If you use your project's code*

module in a larger program, classes in other code modules will not be able to access **internal** *variables and functions.*

You need to be aware that the MSDN documentation list five protection levels for classes. In addition to **public** *and* **internal,** *you could use* **private, protected,** *and* **internal protected.** *In the original version of C#, you could use these three protection levels on any class, but in reality they had no practical use. You still may use these access levels when you define a class within a class, as in the following snippet:*

```
class clsOuter
{
    static public void Main ()
    {
        // Some program statements
    }
    private class clsInner
    {
        // Methods and field in class
    }
}
```

In this code, an instance of **clsInner** *could only be declared within* **clsOuter.**

To declare and define a member function, you must write the code at the point where you declare the function. Unlike C++, there is no separate declaration and definition for functions. To declare and define a member function, write the access keyword for the function followed by the data type the function will return. Next, write the function name, followed by the parameter list. *Do not* write a semicolon after the parameter list.

Immediately after the parameter list, you need to write an opening brace. Some programmers write this symbol on the same line as the function name. Others prefer to write it on a line by itself to make the code block stand out. This book will use the latter as in the following:

```
public double C2F (double fCelsius)
{
    double fFahr = (9 * fCelsius) / 5) + 32;
    return (fFahr);
}
```

Function names may not duplicate a name you used for a variable, but within a class, function names do not have to be unique. The C# compiler recognizes functions by their *signature*, which is a combination of the function name and the parameter list. The signature allows you to *overload* functions, or to write more than one function using the same function name. You will learn about this in Lesson 33, "Overloading Functions and Operators."

Using the Visual Studio Class View Panel

If you are developing your program from within the Visual Studio IDE, you may use the Class View tool window to add variables and functions to your program. Class View displays different forms to add variables and functions, and the various fields on the forms help you to avoid mistakes such as using an invalid access level or mistyping the name of a data type.

To use Class View, you will need to open or create a project in Visual Studio. Create a new project called *F2C*. This will be a simple console application that will convert degrees Fahrenheit to degrees Celsius. Once you start Visual Studio, you have three ways to create a new project:

1 From Visual Studio's opening screen, click on Create New Project on the Start page.

2 Select the File menu New option. Visual Studio, in turn, will display a submenu. Within the submenu, select Project.

3 Simply type **Ctrl+Shift+N.**

All three of these methods will display the New Project dialog box shown in Figure 7.1. Click on Visual C# Projects in the left panel and then click on Console Application in the right panel.

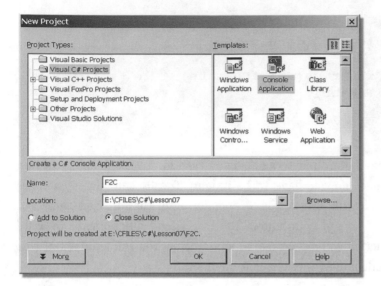

Figure 7.1. To create a new project in the Visual Studio, you need to display the New Project dialog box.

On the bottom half of the New Project dialog box, you will enter the particulars of your project. First, in the Location field, enter the path where you keep your project files. You may use the Browse button to find the directory. The Visual Studio will enter a default path, usually on the C: drive. For performance reasons, I prefer to keep my systems files on one drive, my development tools on another and my project files on yet another drive (with Visual Studio 7.0, you must squeeze all the performance out of your system that you can get).

After you have selected the project path, enter the name of the project, *F2C*.

Just below the Location field, you will see two buttons labeled "Add to Solution" and "Close Solution." Remember that a solution is a workspace, and it may contain more than one project. If you already have a solution open, Visual Studio gives you the option to add any new projects to the solution. For now, however, select the Close Solution button and click on the OK button. Visual Studio will create your project along with basic files that you will need. You should notice that Visual Studio has opened a source-code file, *Class1.cs*, in an editing window. You will refer to this window in the following discussion.

Class View shares the same tool window with the Solution Explorer. Unless you have moved it, the Solution Explorer should be just above the right center of your screen. At the bottom of the tool window, notice that there are two tabs, Solution Explorer and Class View. Click on the Class View tab to display the Class View panel. Expand the list in Class View by clicking on the "+" symbols until your panel looks like Figure 7.2.

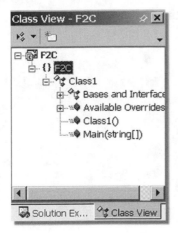

Figure 7.2. The expanded list in the Class View panel shows the elements used in your project.

The first line in Class View in Figure 7.2 shows the project name, *F2C*. This is the *root* level of the tree. If your solution contains multiple projects, the Class View will list each project on a separate line at this level.

The second line is the namespace Visual Studio created for you, which also is *F2C*. As you add namespaces to your project, Visual Studio will list them in the Class View panel. The level of the listing will depend upon where you create a new namespace. You learned that you may nest namespaces, and if you create a new namespace within the *F2C* namespace, the new namespace will appear as a subitem to *F2C*. If, however, you create a new namespace outside the *F2C* namespace, it will appear on the same level.

Select the editing window that contains the *Class1.cs* source file (either click your mouse somewhere in the editing window or click on the Class1.cs tab). Type the following code at the very bottom of the file (make sure you type this code *after* the closing brace for the *F2C* namespace):

```
namespace nsTemperature
{
    class clsTemperature
    {
    }
}
```

Now examine the Class View panel again. The Visual Studio has added a new namespace to the tree, and under that item it has added clsTemperature as a subitem. The IDE is *tracking* your code and updating the tool windows as you add items to your program.

Next, type the same code *inside* the braces that enclose the *F2C* namespace and examine the Class View panel once again. The Visual Studio has added another *nsTemperature* namespace, but this time it is a subitem of the *F2C* namespace. At this point your code should look like the following:

```
namespace F2C
{
    using System;
    namespace nsTemperature
    {
        class clsTemperature
        {
        }
    }

    /// <summary>
    ///     Summary description for Class1.
    /// </summary>
    public class Class1
    {
        public Class1()
        {
            //
            // TODO: Add Constructor Logic here
            //
        }

        public static void Main(string[] args)
        {
            //
            // TODO: Add code to start application here
            //
            return 0;
        }
    }
}

namespace nsTemperature
{
    class clsTemperature
    {
    }
}
```

You should understand that while the two namespaces have the same name, they do not interfere with each other. One namespace is at the *root* level, and the other is nested within the *F2C* namespace. To access the class within the root namespace, you would write *nsTemperature.clsTemperature*. To access the class in the nested namespace, you would write *F2C.nsTemperature.clsTemperature*.

Delete the two namespaces and the classes you added in the previous paragraphs. The Visual Studio should remove them from the Class View as well. You may add classes to your project by typing them into your source-code file, but it often is better to use the Visual Studio tools, which will perform other housekeeping chores for you as you make changes to your project. You will add a class using the tools shortly, but first you need to finish looking over the Class View panel.

Referring to Figure 7.2 again, the third line shows the name of the class that the project wizard created for you, *Class1*. As you add classes to your program, the Class View will list each of them at this level under the namespaces that contain the classes. Below the class name, you will see subitems that contain information about the class. Most classes in C# will contain at least the Bases and Interfaces and the Available Overrides items. You will look at these later. Finally, as subitems, you will see the variables and functions that are members of the class. At this point, *Class1* should contain a constructor function, *Class1()*, and the *Main()* function.

The name *Class1* is not very descriptive, but you have no choice when Visual Studio creates your C# project. I prefer to have my initial class name be related to the purpose of the program. To change the class name, right-click on the name in the Class View and select Properties. The Properties tool window will appear just below Class View. The first line in the Properties tool is (Name). Yes, it is enclosed in parentheses. This is an editable property. Click on the name of the class (Class1) and use the Delete key to erase the name. Type in **clsF2C**.

Examine the name of the class and constructor function in the editing window and the name of the class in Class View. The old name is still in place. Now press the **Enter** key and examine these names again. They all should show the new name, *clsF2C*. This is the best way to change the name of a class. It gives the Visual Studio a chance to change its references from the old name to the new name.

When you must add a class to your program, there is a better way than typing it in as you did earlier. Right-click the mouse on the project name at the top of the Class View panel. From the menu that pops up, select Add and you will get a New Class option. Click on this option to display the Class Wizard dialog box as shown in Figure 7.3.

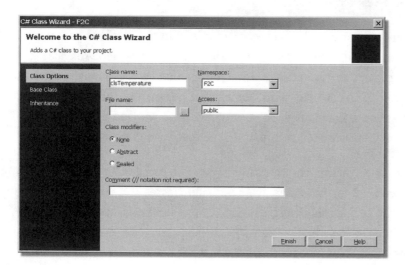

Figure 7.3. The Class Wizard gives you fields and buttons to specify all the information you need to create a new class.

There are three pages to the Class Wizard, which you select by clicking on an item in the large blue area on the left side of the dialog box. In this lesson, you will be concerned only with the first page, Class Options. In the next lesson, you will look at the Base Class and Inheritance pages.

In the Class Name field, type **Temperature**. Notice that the Class Wizard builds the name of the file to hold the class in the File Name field. If you type the name of an existing file, the Wizard will ask if you want to merge the new class into the existing file. Programmers typically keep only one class in a file. This keeps the files from becoming very long and unmanageable. When you compile your program, the C# compiler will collect all the files and create one program file from all of them.

To the right on the Class Name field is the Namespace field. At this point, you have only the *F2C* namespace. You may type the name of another, or a new, namespace in this field, which is the easiest way to create a new namespace, but you must add a class to the namespace at the same time. Again, this is no problem, because in C# a namespace really is useless without a class.

Just below the Namespace field is the Access field. This is not an editable field, and you may select only one of the predefined access keywords listed. The default is *public*. For now you should leave that setting.

Below these fields, you will see a group of buttons labeled Class Modifiers. You will learn about *Abstract* classes in the next lesson, so for now leave the selection on None.

Finally, the Comment field is optional. If you type anything in this field (make it descriptive), the Class Wizard will add the text as a comment using the "//" sequence.

Click the Finish button and the Class Wizard will create a new file and add the class to your project. The new class will contain a single function, the *constructor*. At this point, I usually go to the Properties tool window and add the "cls" prefix to the class name.

Now it is time to add your first variable to a class using the Visual Studio tools. To do this, it is necessary to split the definition of a variable because you will use different wizards to add the different types. Right-click on the *clsF2C* class in the Class View panel and select Add from the pop-up menu. A submenu will appear with four items on it:

❶ Add Method. A method is a function, and you will learn how to add methods to a class later in this section.

❷ Add Property. A property is a member that usually consists of a *private* or *protected* data member (a variable) and functions that you use to assign values to the variable or to retrieve the value of the variable. A property uses *accessors*, functions that operate on the variable that you associate with the property. This is the first distinction C# makes between different types of variables. A property essentially is a "smart" variable that you may access only through its *get* and *set* accessors.

❸ Add Field. A field is an ordinary named variable. Functions may operate on the value of the variable directly without going through accessors. This is the second distinction between variable types and is the one most people associate with the term "variable."

❹ Add Indexer. An indexer enables you to perform array-type operations on classes and structures.

So far in this book, I have used the words "variable" and "property" almost interchangeably to refer to both fields and properties. In C++, a property really is nothing more than a member variable that you access through functions that operate

on the variable. Visual Basic programmers may be more familiar with the C# concept of properties than Visual C++ programmers. In C#, a property usually is, but does not have to be, associated with a member variable. The functions that access a property—the accessors—operate automatically when you assign a value to or retrieve a value from a property.

From this point, I will use "field" or "variable" to refer to an ordinary variable and "property" to refer to a variable that you access through the *get* and *set* accessors.

First, you will add a field (an ordinary variable) to the *clsF2C* class. To do this, you will use the C# Field Wizard. To summon the Field Wizard, summon the menu you just examined above by right-clicking on the class name in Class View, then select Add and then Add Field. The Field Wizard is shown in Figure 7.4.

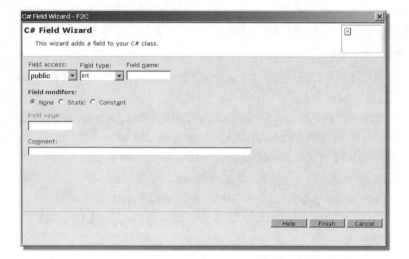

Figure 7.4. You use the C# Field Wizard to add ordinary variables to your classes.

The first box on the wizard is the Field Access. This is not an editable box, and you may select only one of the five predefined access keywords in the drop-down box. The default setting is *public*.

The Field Type is an editable area that contains the C# fundamental data types in a drop-down box. You may select one of the fundamental types, or enter a user-defined type such as a class name or structure name.

The Field Name is where you will enter the name of your variable. In Visual C++, it is common practice to prefix the name with "m_" (such as *m_iVar*) to indicate the variable is a member of a class rather than a variable you declare in a function. That convention has not taken root in C#, however.

Leave the Field Modifier set to None for now. Later in this lesson, you will learn about *static* variables and functions. You learned about *constant* variables in Lesson 6, "Using Variables to Store Information In C# Programs."

The wizard will disable the next box, the Field Value, unless you select Constant as the field modifier. As you learned in Lesson 6, you must assign a value to a *const* variable when you declare the variable. You can set that value here.

If you add text to the Comment box, the wizard will add a comment line above the declaration containing the text you enter here. You do not need to include the "//" sequence. You may leave the Comment box blank.

Type **m_iTest** in the Field Name and type **Test field** in the Comment box. Press the **Finish** button and examine the *clsF2C* class. The wizard has added the following lines to your code:

```
/// Test field
public int m_iTest;
```

In addition, the variable name now appears in the Class View panel under the clsF2C class name.

Adding a property to a class requires a little more forethought. You must consider how you will use the property to decide which accessors to include. You will make this decision on the C# Property Wizard. To summon the Property Wizard, display the menu as you did earlier, except this time select Add Property. The Property Wizard is shown in Figure 7.5.

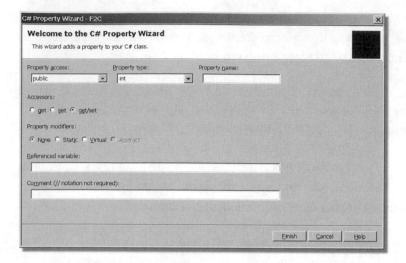

Figure 7.5. The C# Property Wizard is the best method to add variables to your program.

The first three boxes—Property Access, Property Type and Property Name—are the same as for the Field Wizard.

To select Accessors, you must understand what the accessors do. The *get* accessor is a block of code that your program will execute when you read the value of the property. It always returns a value of the property's data type, but your program may execute other statements before returning a value. If you must permit functions to retrieve the value of the property, then you must include the *get* accessor.

The *set* accessor is a block of code your program will execute when you assign a value to the property. Within the *set* accessor, there is an *intrinsic* variable named *value*, which contains the value you assign to the property. (The variable is intrinsic because you do not have to declare the variable to use it.) You may use *value* to set another variable. If you do not save it in another variable, the value of *value* is lost when the *set* accessor terminates. If you will permit functions to assign a value to the property, you must include the *set* accessor.

Normally, your program will use both the *get* and *set* accessors, so you should select the button labeled "get/set."

The Property Modifiers buttons also will depend upon how you plan to use the property. You will learn about *static* variables and properties later in this lesson. You will learn about the *virtual* modifier in Lesson 8. The wizard will disable the Abstract button unless you are adding the property to an *abstract* class, which also will have to wait until Lesson 8.

It is important that you understand at this point that a property itself does not store a value. It simply operates on values through the *get* and *set* accessors. If you must save the value assigned to a property (and you usually do), you must set aside another variable to hold the value. To do this, you must add a field to your class. The field should be the same data type as the property. For example, if the data type of the property is *long*, the data type of the referenced variable should be *long*.

An example will make this easier to understand. Add a property to the *clsF2C* class. Select *double* as the Property Type. Name the property *m_fTemp*. Type **Test property** in the Comment box. Click your mouse on the Finish button to add the property to the class. Examine the *Class1.cs* edit window and you will see that the wizard has added the following code to your program:

```
/// Test property
    public double m_fTemp
    {
        get
        {

            return 0;
        }
        set
        {
        }
    }
```

You now must add a private or protected field with the same data type to your class, as shown here:

```
    protected double m_fFahr;
```

The curious construction makes *m_fTemp* look almost like a variable and almost like a function, but not really either. When you see this construction in other code, you should recognize it as a property.

Modify the *Main()* function to set and display the value of the *m_fTemp* property, then add code to the *get* and *set* accessors to save and retrieve the value in *m_fFahr*. The *Class1.cs* file should look like the following:

```csharp
using System;

namespace F2C
{
    /// <summary>
    ///     Summary description for Class1.
    /// </summary>
    public class clsF2C
    {
        public clsF2C()
        {
            //
            // TODO: Add Constructor Logic here
            //
        }

        public static int Main(string[] args)
        {
            clsF2C temp = new clsF2C();
            temp.m_fTemp = 20;
            //
            // TODO: Add code to start application here
            //
            Console.WriteLine("The value of m_fTemp is {0}",
                            temp.m_fTemp);
            Console.WriteLine("The value of m_fFahr is {0}",
                            temp.m_fFahr);
            return 0;
        }

        /// Test field variable
        public int m_iTest;

        /// The value assigned to m_fTemp will be a
        /// Celsius temperature. The value returned
        /// will be the corresponding Fahrenheit value.
        public double m_fTemp
        {
            get
            {
                return (m_fFahr);
```

```
            }
        set
        {
// value is an intrinsic variable in the set accessor
            m_fFahr = (9 * value) / 5 + 32;
        }
    }

        protected double m_fFahr;
    }
}
```

When you run the program, you will see that, although *Main()* assigns a value of 20 to *m_fTemp*, when the program displays the value on the console, it displays 68, the Fahrenheit equivalent of 20 degrees Celsius. By using a property instead of an ordinary variable (field), the conversion is automatic. You assign one value and you pull out another.

This has been a long section, but you still must understand how to add functions, or methods, to your classes. To do this, you will use the C# Method Wizard. Select the menu as you did earlier in this section and select Add Method to start the Method Wizard as shown in Figure 7.6

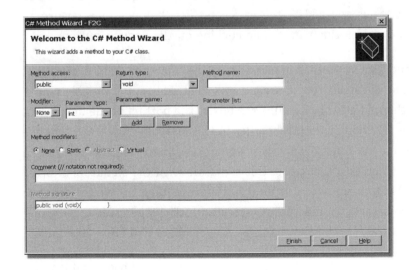

Figure 7.6. You will use the C# Method Wizard to add functions to a class.

You assign the access level, the function return type, and the function name in the first row of boxes. Return Type is an editable box, and you may select any of the fundamental data types from the drop-down box or type in a user-defined type.

You should note that the Return Type box may not be empty, and there is no entry in the drop-down box to omit a return type. This means that you may not use the Method Wizard to add additional constructors to your class (a constructor function may not have a return type). If you want to add constructors, you will have to enter them manually.

You build the function's parameter list one parameter at a time using the boxes in the second row. The Modifier box contains a drop-down list for setting *reference* and *out* parameters. You will learn more about these in Lesson 24, "Changing Parameter Values within a Function." Leave this box set to None for now.

The Parameter Type is the data type of the parameter. You may select from one of the fundamental types in the drop-down box, or enter a user-defined type.

Enter in the Parameter Name box the identifier that the function will use to access the parameter. The identifier may be any name you want as long as it begins with an underscore character or an alpha character ("a" to "z" or "A" to "Z") and does not contain any spaces or C# operator symbols. The name *does not* have to be the same as another variable in the class.

When you have selected the modifier and type and entered a name for the parameter, click on the Add button. The wizard will add the parameter and pertinent information to the Parameter List box. The parameters will appear in the function definition in the same order they appear in the Parameter List. You cannot edit the items you add to the parameter list, but you may remove an item by selecting it with the mouse, then clicking on the Remove button.

Type **MyMethod** in the Method Name box in the first row. Add a couple of parameters using the fields in the second row. First add an *int* variable named *iVar* and click the Add button. Examine the box at the bottom of the dialog box, the one labeled "Method signature." Earlier in this lesson, you learned that the C# compiler recognizes functions by their signatures. When you add a method to your class, the compiler will compare this signature with the signature of other member functions to make sure it is unique. Unfortunately, the wizard does not actually use the signature, and you may create functions with identical signatures using the wizard. When that happens, the compiler will issue an error.

The Method Modifiers are the same as for the Property Wizard page, and you will learn about the options later. Do not select any of these buttons for now.

Finally, you may add comments to the definition by entering them in the Comments box, the same as you did for the Properties Wizard. You do not need to enter the "//" sequence.

Click the Finish button and the wizard will add the *MyMethod()* to your class. The *Class1cs* file now should look like the following:

```csharp
using System;

namespace F2C
{
    /// <summary>
    ///     Summary description for Class1.
    /// </summary>
    public class clsF2C
    {
        public clsF2C()
        {
            //
            // TODO: Add Constructor Logic here
            //
        }

        public static int Main(string[] args)
        {
            clsF2C temp = new clsF2C();
            temp.m_fTemp = 20;
            //
            // TODO: Add code to start application here
            //
            Console.WriteLine("The value of m_fTemp is {0}",
                            temp.m_fTemp);
            Console.WriteLine("The value of m_fFahr is {0}",
                            temp.m_fFahr);
            return 0;
        }

        /// Test field variable
        public int m_iTest;
```

```
        /// The value assigned to m_fTemp will be a
        /// Celsius temperature. The value returned
        /// will be the corresponding Fahrenheit value.
        public double m_fTemp
        {
            get
            {
                return (m_fFahr);
            }
            set
            {
// value is an intrinsic variable in the set accessor
                m_fFahr = (9 * value) / 5 + 32;
            }
        }

        protected double m_fFahr;

        public void MyMethod (int iVar)
        {
        }
    }
}
```

The wizard adds an empty method to your class. The wizard has no way of knowing what purpose you intend the method to perform, so adding code to the method is your responsibility.

Understanding the *this* Variable

In the last section, you met your first *intrinsic* variable in the *set* accessor for properties. An intrinsic variable is a built-in variable, one that you do not need to declare before using.

Sometimes a class needs to refer to itself. In C#, classes and structures inherit an intrinsic variable called *this*. You don't need to declare a *this* variable; it is a part of the class itself, and you cannot change its value. In addition, *this* may be used only within member functions, but you may use it in calls to other functions where you must identify the class *instance*. Each instance of a class has its own *this* variable.

You may use the *this* variable to access class members and to resolve ambiguity between class member variables and variables that you declare in a function. The *this* variable is very useful in C++, where its value points to the memory location of the class instance. Your C# statements, however, will rarely use memory addresses directly, making the *this* variable less useful. You should learn how to use the variable, however.

Suppose you have a class named *clsEmployee* that contains members to hold the employee's name and identification number. The parameters to the constructor have the same name as the class member variables. In this case, you may use the *this* variable to resolve the ambiguity, as shown in the following snippet:

```
class clsEmployee
{
    clsEmployee (string name, int ID)
    {
        this.name = name;
        this.ID = ID;
    }
    string name;
    int ID;
}
```

In this snippet, *this.name* and *this.ID* always refer to the class member variables. If you tried to write an assignment such as *name = name* in the constructor, you would only assign the parameter to itself.

Understanding the *static* Modifier for Class Members

Throughout this lesson, you have seen references to a *static* modifier for member variables and methods. When you declare a method or variable *static*, it takes on some special properties, and has some special restrictions.

When you create a variable from a class definition, you are creating an *instance* of the class. The compiler adds code to your program to set aside memory for the class members. You then may access the members using the variable name you assigned to the instance. In the following snippet, you create an instance of a class named *clsEmployee*, then access the *name* variable using the instance name:

```
class clsMain
{
    public static void Main()
    {
        clsEmployee Worker = new clsEmployee();
        string strName = Worker.name;
    }
}
```

Each time you declare a new instance of the class, the program creates new variables for the new instance. Changing the value of a variable in one instance does not affect the value in another instance. You cannot access a member variable before you create a class instance because the variable simply does not exist yet.

When used this way, programmers say the members "belong to the instance" of the class. This is true of all ordinary variables and functions that are members of a class.

When you use the *static* modifier on a variable, however, the program does not create a new variable for each instance of the class. Instead, the program creates one of each *static* variable, and all instances of a class must share the same variable. Changing the value of a *static* variable in one instance changes the value for all instances. Within a Worker class, for example, you might use a *static* variable to hold information that is company-specific (such as the company phone number), so that when the application changes the value, the change automatically appears in each employee's information.

Programmers say that a *static* member variable "belongs to the class" rather than the instance. The following program, *Static.cs*, shows how two instances of a class share the same *static* variable:

```
/*
    Static.cs. Demonstrate the use of static variable
             members of a class.

    Compile this program using the following command line:
             csc StrForm.cs
*/

namespace nsStatic
{
    using System;

    class clsStatic
    {
        static void Main()
        {
            clsTest First = new clsTest ();
            clsTest Second = new clsTest ();
            First.SetVal (42);
            First.Write();
            Second.Write();
        }
    }

    class clsTest
    {
        public static int iVar;

        public void SetVal (int val)
        {
            iVar = val;
        }
        public void Write()
        {
            Console.WriteLine ("iVar = {0}", iVar);
        }
    }
}
```

Although you set the value of *iVar* in the *clsTest* class using only the first instance of the class, the value also changed for the second instance.

You do not need to assign an initial value to a *static* member variable. The compiler initializes all *static* members to 0, but the compiler also may issue a warning that you did not give the variable a value.

Like member variables, you may not access a member function until you create an instance of a class. Because member functions may operate on member variables, allowing you to call a member function without an instance of the class might result in your functions operating on variables that do not yet exist.

However, when you declare a member function *static*, you may call the function even before you create an instance of the class. Obviously, this means that *static* functions may not access ordinary member variables that might not exist yet. However, *static* functions may access *static* member variables because your program creates *static* variables when you load the program in memory.

You have seen the *Main()* function as a member of a class. Your program must call *Main()* to run the program and, perhaps, create instances of a class. But if *Main()* were an ordinary function, you would not be able to call it until you create an instance of the class. By declaring it *static*, you resolve this problem.

WHAT YOU MUST KNOW

You have covered a lot of ground in this lesson. You learned how to define and use classes in your program. You learned how to add member variables, methods and properties using the Class View tool window. You also learned the difference between field variables and properties, and how to use the *get* and *set* accessors for properties. In addition, you learned about the *this* intrinsic variable that your classes may use and how to declare and use *static* variables and functions. In Lesson 8, "Object-Oriented Programming and C#," you will learn how to use a class as a *base class*, and how to derive a new class to inherit the properties, variables and methods of the base class. You also will learn about the object-oriented concept of *polymorphism*. Before you continue with Lesson 8, however, make sure you have learned the following key concepts:

- The class is the basic unit for object-oriented programming. In C#, classes contain all the variables and functions that your program uses.

- You may define as many classes as your program needs. The name of a class must be unique within the namespace, but classes in different namespaces may have the same name.

- The Class View tool window is the access point for wizards that help you to add variables, properties and methods to your classes.

- A *field* is an ordinary member variable of a class. You may add a field using the C# Field Wizard.

- A *property* is a variable that you access through *get* and *set* accessors. A property does not necessarily store a value. You may add properties to a class using the C# Property Wizard.

- A *method* is a function that is a member of a class. The C# Method Wizard assists you in adding methods to your classes.

- Each instance of a class object inherits an intrinsic variable named *this* that the class instance may use to identify itself.

- Member functions and variables that you declare *static* have special properties, but the C# language also imposes special restrictions on these members.

OBJECT-ORIENTED PROGRAMMING AND C#

*i*n the last lesson, you learned about C# classes and how to add fields, properties and methods to a class, and how the class is the primary element for *object-oriented programming* with C#. Object-oriented programming is built on three primary concepts: encapsulation, inheritance and polymorphism. C# enforces encapsulation by requiring that all variables and functions be members of a class. Inheritance involves using an object definition that you already have written to *derive* a new object that assumes the variables, properties and functions of the original object. Polymorphism refers to the ability of a derived object to change forms by modifying the behavior of its parent, or base, object. In this lesson, you will learn how to use these concepts in your programming and how to write base classes from which you will derive new classes. By the time you finish this lesson, you will understand the following key concepts:

• Ordinary variables contain information only about the *value* of the variable. Objects may contain information such as properties and methods that set and retrieve values.

• C# is an object-oriented programming language. The class is the basic mechanism of object-oriented programming and your C# program must contain at least one class definition.

• C# variables are *value*-types or *reference*-types. Ordinary variables that contain only value information are value-types. Objects, such as those you create from a class definition, are reference-types.

- Encapsulation is the process of packaging related information into an object. C# requires that all variables, properties and methods be members of a class, and so enforces the idea of encapsulation.
- Inheritance refers to the ability to build upon objects you already have defined to create a new object that assumes the properties of its parent.
- Using polymorphism, you may inherit the properties of a parent object and modify the way the parent object works.
- C# implements polymorphism through the use of the *virtual* keyword, which allows a method or property in a derived object to override the property or method in a base class.

Understanding Objects

Not long ago all programming was task- or procedure-oriented. In this paradigm, the computer executed one statement of the program then moved on to the next. You could trace the flow of the program simply by looking at the sequence of statements in a program. A variable was a memory location in which the computer stored a value. When the computer encountered a subroutine or function call in the program, it simply transferred control to the subroutine.

In object-oriented programming, the flow becomes more complex. A variable constructed from an object definition might contain not one but many values, and the object might contain other code to operate on those values. Simply declaring an object in a program might invoke other code, making it more difficult to trace the flow (but not impossible).

Variables and objects are distinctly different animals in object-oriented programming. A variable contains only the value that has been stored in a memory location. An object may contain methods or functions to store, retrieve or manipulate one or more values.

When you define a class object in C#, you are not creating any of the variables or methods that are part of the class. The class definition gives the compiler a *template* that it will use to set aside memory to hold the class members when you declare an *instance* of the class. An instance is a variable declaration that actually creates the object. Except for *static* members, none of the members of a class exist until you create an instance.

Using Value- and Reference-Types

The C# language divides data types into two distinct types, *value-types* and *reference-types*. Because of the way C# handles values, the difference is not as great as you might suspect. The major difference between the two types is that you can access value-types directly, but you must access reference-types indirectly through a *pointer*, although C# hides the details of the pointer from you.

The value-types include the fundamental data types such as *int, long,* or *double*. The exception is the *string* data type, which, although C# lists it as a fundamental data type, really is an alias for the *String* class. The functions and overloaded operators in the *String* class give the *string* data type the feel of a fundamental type.

An ordinary variable—a *field*—is a value-type. You have learned that a variable is a memory location that contains the value that you have assigned to the variable. When you deal with a variable, you really are using that value rather than directly accessing the memory address. For example, when you use a variable as an argument to a function call, C# makes a *copy* of the value and uses that in the function call rather than the memory address itself. If the called function changes the value of the argument, it does not change the original value of the variable. The following program, *ValType.cs*, shows that a variable remains unchanged when you use it in a function call:

```
namespace nsValue
{
    using System;
    class clsMain
    {
        public static void Main()
        {
            clsMain ValTest = new clsMain();
            int value = 42;
            Console.WriteLine ("In Main, value = {0}\n",
                                value);
            ValTest.ModifyValue (value);
            Console.Write ("In Main, value still = {0}\n",
                            value);
```

```
        }

    public void ModifyValue (int value)
    {
        value = 21;
        Console.Write("In ModifyValue, value = {0}\n",
                      value);
    }
  }
}
```

When you compile and run *ValTest.cs*, you will see the following output:

```
In Main, value = 42
In ModifyValue, value = 21
In Main, value still = 42
```

Although you passed the variable *value* as an argument to *ModifyValue()* and even used the same name for the argument in the function, the actual value of the variable in *Main()* remains unchanged. *ModifyValue()* does not "know" the memory address of the original variable, and thus cannot modify the original value.

Other value-types include *structures*, which you will learn about in Lesson 14, "Using Structures in C# to Group Related Data," and the enumerated values, which you will cover in the next topic of this lesson.

Reference-types, on the other hand, operate indirectly using the memory address of the variable or object. You may change a value-type to a reference-type by using the *ref* keyword when you used a value-type variable, as in the following program, *RefVar.cs*. Notice that this is the same program as *ValType.cs* except the *ref* keyword appears in the function call and in the function definition:

```
/*
    RefVar.cs  — Passes the address of a variable through a ref
                    type rather than a value-type

      Compile this program with the following
              command line:
                  csc RefVar.cs
 */
namespace nsValue
{
    using System;
    class clsMain
    {
        public static void Main()
        {
            clsMain ValTest = new clsMain();
            int value = 42;
            Console.Write ("In Main, value = {0}\n",
                            value);
            ValTest.ModifyValue (ref value);
            Console.Write ("In Main, value now = {0}\n",
                            value);
        }

        public void ModifyValue (ref int value)
        {
            value = 21;
            Console.Write("In ModifyValue, value = {0}\n",
                            value);
        }
    }
}
```

When you run this program, you see the following output on your screen, showing that the function call did indeed change the value of the original variable.

```
In Main, value = 42
In ModifyValue, value = 21
In Main, value now = 21
```

In C and C++, you could accomplish this by passing the function a *pointer* to the variable, which is the actual memory address of the variable. Except under certain conditions, however, C# does not permit you to use pointers. Using the *ref* keyword accomplishes the same result, however.

Objects such as instances of a class always are *reference-type* variables. You must create instances of a class using the *new* operator. In the previous example, you created an instance of the *clsMain* class using the following declaration:

```
clsMain ValTest = new clsMain();
```

The *new* operator actually returns a pointer to the memory location where the class instance is stored. However, C# calls the resulting variable a *reference* variable. You really are using pointers without going through the process of declaring and storing pointers. The difference is more semantic than real.

Declaring and Using Enumerated Values

Enumerated values really are nothing but convenient aliases for values or a group of values. An enumerated value is not really a variable, but a value that you reference by name. You cannot change the value of an enumerated type as you can with a variable. Enumerated values must be integer types such as *int*, *long*, or *byte*. You cannot use a floating point value such as *double* or a user-defined data type as an enumerated value.

To declare an enumerated value, type the keyword *enum* followed by the identifier for the enumerated list. Then type an open brace. Type the alias for the first enumerated value. End the enumeration sequence with a closing brace and a semicolon. The following creates an enumerated list that contains only one value:

```
enum Vals
{
    Zero
};
```

After this declaration, you may use *Vals.Zero* as an alias for the value *0*.

You may create several aliases in a single enumeration by separating the names by a comma. In this case, each successive name assumes the next integer value (the same as adding 1 to the previous value):

```
enum Vals
{
    Zero, One, Two, Three, Four
};
```

In this statement, *Vals.Zero* takes on the value 0, *Vals.One* the value 1 and so on. Enumerated values always start counting at 0 unless you assign a different starting value. The following declaration assigns the same values to the identifiers except the *Zero* identifier is missing:

```
enum Vals
{
    One = 1, Two, Three, Four
};
```

You also may assign each identifier in an enumeration a separate value. The following enumeration creates aliases for the first five even numbers:

```
enum Even
{
    Two = 2, Four = 4, Six = 6, Eight = 8, Ten = 10
};
```

Enumerations are handy for isolating your code from value changes. For example, suppose your program deals with days of the week as integer values with Sunday having the value 0. You write lots of code using numbers from 0 to 6 to represent the days of the week. You could, instead, use enumerated values:

```
enum Weekdays
{
    Sunday, Monday, Tuesday, Wednesday,
    Thursday, Friday, Saturday
};
```

Now you use the aliases for the values of the days of the week. If you have to change the base, say, start using 1 for Sunday because that is the way a remote system you connect with handles days of the week, you simply return to the enumeration statement and change the starting value:

```
enum Weekdays
{
    Sunday = 1, Monday, Tuesday, Wednesday,
    Thursday, Friday, Saturday
};
```

Now the correction will be applied throughout your program when you recompile your code.

Although enumerations are integer-type values, in C# you must cast them to the specific data type of your variable:

```
int Today = (int) Weekdays.Tuesday;
```

If you do not include the cast, the C# compiler will generate an error and will not compile your program.

If you declare an enumeration in a class definition, the enumeration belongs to the class, and you may not use it outside of the class. However, because enumerations do not actually generate any memory storage themselves, you may declare an enumeration outside of a class definition. The following code fragment, for example, creates the enumerated type *Days* within the nsEnums namespace. Then, within the class *clsMain*, the code creates a second enumerated type *Weekdays*. When you define an enumerated type within a class, only that class has knowledge of the type:

```
namespace nsEnums
{
    //   the following enumeration may be used by any
    //   class in the namespace:
    enum Days
    {
        Sun, Mon, Tue, Wed, Thu, Fri, Sat
    }

    class clsMain
    {
        // The following enumeration may be used only within
        // the clsMain class:
        enum Weekdays
        {
            Sun, Mon, Tue, Wed, Thu, Fri, Sat
        }
    }
}
```

Understanding Inheritance

Inheritance is a fundamental principle in object-oriented programming. Using inheritance, you may use an existing class as a *base* class to derive a new class. The new class will inherit the member variables and functions of the base class.

To derive a new class from an existing class, write the new class definition as you normally would. However, after the class name, write a colon followed by the name of the base class as shown below (assuming that you already have defined *clsBase*):

```
class clsNew : clsBase
{
// New class members
}
```

When you derive one class from another, the new class is a *derived* class and becomes a new user-defined data type. Generally, you use a more generic class as the base class, leaving the derived class to provide the more specific details. For example, suppose you wanted to build a couple of classes to describe plants and animals. These groups share one common trait: They are life forms. You could build a base class called *clsLifeform* to hold the common information, then derive new classes for plants and animals. The following program, *Inherit.cs*, shows how this can be done:

```
namespace nsInherit
{

    using System;

    enum Kingdoms
    {
        Animal, Plant
    };

    class clsMain
    {
        public static void Main()
        {
            clsAnimal Animal = new clsAnimal ();
            clsPlant Plant = new clsPlant ();
            Animal.Show();
            Plant.Show();
        }

    }

    class clsLifeform
    {
        protected int Kingdom;
    }
```

```
class clsAnimal : clsLifeform
{
    protected int Phylum = 0;

    public clsAnimal ()
    {
        Kingdom = (int) Kingdoms.Animal;
    }
    public void Show()
    {
        Console.Write ("This is an animal life form\n");
        Console.Write ("Kingdom is {0}\n", Kingdom);
    }
}

class clsPlant : clsLifeform
{
    public clsPlant ()
    {
        Kingdom = (int) Kingdoms.Plant;
    }
    public void Show()
    {
        Console.Write ("This is a plant life form\n");
        Console.Write ("Kingdom is {0}\n", Kingdom);
    }
}
}
```

In the preceding code, notice that the *Kingdom* variable is a member of the *clsLifeform* class, yet you access the variable as though it were a member of each of the derived classes. Because each class inherits the *Kingdom* variable, you do not have to declare the *Kingdom* variable in each derived class.

After you have defined the derived class, you then may use it as a base class. In this case, the new derived class would inherit the properties of both classes. Continuing with the *clsLifeform* class, you could derive a *clsProtozoa* class, and then use *clsProtozoa* to derive yet another class, *clsAmoeba* as in the following program, *Inherit2.cs*:

```
namespace nsInherit
{

    using System;

    enum Kingdoms
    {
        Animal, Plant
    };

    class clsMain
    {
        public static void Main()
        {
            clsAnimal Animal = new clsAnimal ();
            clsPlant Plant = new clsPlant ();
            clsProtozoa Protozoa = new clsProtozoa ();
            clsAmoeba Amoeba = new clsAmoeba ();
            Animal.Show();
            Plant.Show();
            Protozoa.ShowProtozoa();
            Amoeba.ShowAmoeba();
        }


    }

    class clsLifeform
    {
        protected int Kingdom;
    }

    class clsAnimal : clsLifeform
    {
        protected int Phylum = 0;

        public clsAnimal ()
        {
            Kingdom = (int) Kingdoms.Animal;
        }
        public virtual void Show()
        {
            Console.Write ("This is an animal life form\n");
```

```csharp
            Console.Write ("Kingdom is {0}\n", Kingdom);
        }
    }

    class clsPlant : clsLifeform
    {
        public clsPlant ()
        {
            Kingdom = (int) Kingdoms.Plant;
        }
        public virtual void Show()
        {
            Console.Write ("This is a plant life form\n");
            Console.Write ("Kingdom is {0}\n", Kingdom);
        }
    }
    class clsProtozoa : clsAnimal
    {
        public clsProtozoa ()
        {
            Kingdom = (int) Kingdoms.Plant;
        }
        public void ShowProtozoa()
        {
           Console.Write("This is a protozoan life form\n");
           Console.Write ("Kingdom is {0}\n", Kingdom);
        }
    }
    class clsAmoeba : clsProtozoa
    {
        public clsAmoeba ()
        {
            Kingdom = (int) Kingdoms.Plant;
        }
        public void ShowAmoeba()
        {
            Console.Write ("This is an amoeba life form\n");
            Console.Write ("Kingdom is {0}\n", Kingdom);
        }
    }
}
```

Each derived class continues to inherit the *Kingdom* variable that you declared in the *clsLifeform* class. Notice, however, that each constructor in the inheritance chain writes its own version into the value of *Kingdom*. In Lesson 10, "Understanding Constructors and Destructors," you will learn how to pass values from one constructor to another in an inheritance chain.

You should note also that the *Kingdom* variable has an access level of *protected*. Inheriting from a class does not change the access of member fields, properties or methods. If a member of a base is *private*, derived classes may not access the base class member. Derived classes do have access to *protected* members, however.

Examining Encapsulation and Visibility

One of the characteristics of an object is that it may contain all the fields, properties and methods that it needs to describe itself. This is the object-oriented programming concept of *encapsulation*.

In C#, as you have learned, all variables and methods must be members of a class. In this way, C# enforces the principle of encapsulation.

Encapsulation also means that when you create a class object variable, you create all the member fields and properties at the same time. You can access the members only through the class object, so when your code is finished with the object and it gets destroyed, the members get destroyed at the same time.

The range of statements through which you may access the members of a class is the *scope* of the class object. If you create an object within a function, you may access the object and its members only while your code is executing statements within the function. Because you cannot use an object until you declare it, the scope then is the statement where you declare the object to the end of the function.

You will learn more about scope in Lesson 11, "Understanding Class Scope and Access Control."

Understanding Polymorphism

Through the principle of inheritance, one class can inherit and use the properties of another class. In a hierarchy of classes such as the life forms classes in the previous examples, the first base class defined some very basic behavior. The next group then became more specific, splitting life forms into the animals group and the plants group.

One common element is that each of the derived classes could print a description of itself, but the functions had to have different names to avoid getting warnings from the compiler. By changing the design slightly, you could provide a common function name to display the description and let the program select the proper function to call. Your derived classes would thus inherit the properties and methods of their parent classes, but with some changes to meet the needs of the derived class.

This is the principle of *polymorphism*. The primary mechanism through which C# classes support polymorphism is the *virtual* function. A function with the same name in a derived class then may provide its own version of the virtual function by using the *override* keyword.

To demonstrate this in the life forms class hierarchy, you need to redesign the *clsLifeform* class slightly. You need to add a new function, *Display()*, that does nothing but call the *Show()* function. Then add the *Show()* function to the *clsLifeform* class using the *virtual* keyword.

Next, in each of the derived classes, change the function that writes a description to the screen to *Show()*, but add the *override* keyword to each.

In the *clsMain* class, change the lines that display the information to call the *Display()* function in the *clsLifeform* class. (Excuse any biological inaccuracies; the intent is to show how a base class can call a function in a derived class, although the base class had no knowledge of the derived class when you wrote it). When you have done this, your code should look like the following program, *Virtual.cs*:

```
/*
    Virtual.cs  — demonstrates polymorphism using virtual
                    functions
    Compile this program with the following
            command line:
                C:> csc Virtual.cs
 */


namespace nsInherit
{

    using System;

    enum Kingdoms
    {
        None, Monara, Fungi, Protista, Plantae, Animalia
    };

    class clsMain
    {
        public static void Main()
        {
//  If you uncomment the next line and attempt to declare
//  an instance of clsLifeform, the compiler will issue
//  an error.
//              clsLifeform Lifeform = new clsLifeform ();
            clsAnimal Animal = new clsAnimal ();
            clsPlant Plant = new clsPlant ();
            clsProtozoa Protozoa = new clsProtozoa ();
            clsAmoeba Amoeba = new clsAmoeba ();
            Animal.Display();
            Plant.Display();
            Protozoa.Display();
            Amoeba.Display();
        }

    }

    abstract class clsLifeform
    {
```

```csharp
        protected int Kingdom
        {
            get
            {
                return (m_Kingdom);
            }
            set
            {
                m_Kingdom = value;
            }
        }
        private string[] strKingdoms = {"", "Monara",
                                "Fungi", "Protista",
                                "Plantae", "Animalia"};

        public string strKingdom
        {
            get
            {
                return (strKingdoms[m_Kingdom]);
            }
        }

        private int m_Kingdom = 0;

        protected string strName
        {
            get
            {
                return (m_Name);
            }
            set
            {
                m_Name = value;
            }
        }
        private string m_Name = "";

        public virtual void Move ()
        {
        }
        public void Display ()
```

```
        {
            Move ();
        }
}
class clsPlant : clsLifeform
{
    public clsPlant ()
    {
        Kingdom = (int) Kingdoms.Plantae;
    }
    public override void Move()
    {
        Console.Write ("Members of the {0} kingdom ",
                        strKingdom);
        Console.Write ("Have limited movement\n\n");
    }
}
class clsAnimal : clsLifeform
{
    public clsAnimal ()
    {
        Kingdom = (int) Kingdoms.Animalia;
        strName = "Animals";
    }
    public override void Move()
    {
        Console.Write ("{0} are members of the " +
                        "{1} kingdom\n",
                        strName, strKingdom);
        Console.Write ("They are multicellular and " +
                        "usually move freely about " +
                        "their environment\n\n");
    }
}
class clsProtozoa : clsLifeform
{
    public clsProtozoa ()
    {
        Kingdom = (int) Kingdoms.Protista;
        strName = "Protozoa";
    }
    public override void Move()
```

```
        {
            Console.Write ("The {0} is a member of the " +
                             "{1} kingdom\n",
                             strName, strKingdom);
            Console.Write ("It can move about in its " +
                             "environment\n\n");
        }
    }
}
class clsAmoeba : clsProtozoa
{
    public clsAmoeba ()
    {
        Kingdom = (int) Kingdoms.Animalia;
        strName = "Amoeba";
    }
    public override void Move()
    {
        Console.Write ("The {0} is a member of " +
                         "the {1} kingdom.\n",
                          strName.ToString(),
                          strKingdom.ToString());
        Console.Write ("It moves by extending " +
                         "protoplasmic \"legs\"\n\n");
    }
}
}
```

In *clsMain*, each class object now calls the same function, *Display()* in the *clsLifeform* base class. Then *Display()* simply calls the *Show()* function. Although *Move()* in the base class is empty—it has no statements—your program still prints the proper information.

When you use the *virtual* and *override* keywords, your program will search through the derived classes for any overridden function with the same name. It will continue searching to the most distant ancestor, then call this function.

Actually, your program decides which function to call when you first load the program. Instead of providing the program with the functions to call, the compiler prepares a table of *virtual* functions in your program. The decision on which function to call is deferred until you actually run the program.

Occasionally you may see a reference to "sealed" classes in C#. When you used the sealed keyword on a class definition, it marks the end of an inheritance chain. You cannot derive a new class from a sealed class, nor may you use the sealed and abstract keywords together for a class definition.

C# Does Not Allow Multiple Inheritance

An important difference between C# and C++ is that C# does not allow a derived class to inherit variables and methods from more than one base class. If you have studied C++, you may have run across instances where you need to combine the characteristics of more than one hereditary branch. Ultimately, both branches in this "multiple inheritance" would lead back to the same ancestor class, as shown in Figure 8.1.

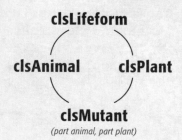

Figure 8.1 Using multiple inheritance to create an object.

*Here, **clsMutant** inherits two copies of **clsLifeform**, one through **clsAnimal** and another through **clsPlant**. This leads to an ambiguous reference—from which copy does **clsMutant** inherit variables and methods?*

*C++ contains mechanisms to resolve this ambiguity. C# simply does not allow this type of multiple inheritance and thus avoids the ambiguous reference. However, C# does allow multiple inheritance of **interfaces**.*

Properties also may be *virtual*. If you must override the *Kingdom* variable in derived classes (perhaps to provide phylum information), you can define *Kingdom* as a *virtual* property as follows:

```
protected virtual int Kingdom
{
    get
    {
        return (m_Kingdom);
    }
    set
    {
        m_Kingdom = value;
    }
}
private int m_Kingdom;
```

In a derived class, you would provide your own property with its own *get* and *set* accessors using the *override* keyword:

```
protected override virtual int Kingdom
{
    get
    {
        return (m_Kingdom);
    }
    set
    {
        m_Kingdom = value;
    }
}
private int m_Kingdom;
```

You should notice that in this example you would have to provide your own variable in which to save the value. The *m_Kingdom* field in the base class is *private*, and derived classes may not access it. However, you may override a field in a base class simply by providing a field with the same name in the derived class.

Using Abstract Classes

Sometimes you will need a class to serve only as a base class. In the previous example, *clsLifeform* is nothing but a container for the *Kingdom* property and the *Display()* and *Show()* functions. It contains no code to display information. However, there is nothing to keep you from declaring an instance of *clsLifeform* and then calling its *Display()* method:

```
clsLifeform Lifeform = new clsLifeform();
Lifeform.Display();
```

Of course, this would get you nowhere. Your program would not even display a blank line when you call the *Display()* method.

To avoid this situation, you may declare *clsLifeform* as an *abstract* class, as shown in the sample code in the previous topic. Then, when you try to declare an instance of the class, the compiler will issue an error that you "cannot create an instance of the abstract class." Simply add the *abstract* keyword to the *clsLifeform* definition:

```
abstract class clsLifeform
```

Even worse, suppose you forget to override the *Show()* method in your derived class. When you call the *Display()* method, your program again would display nothing because it would call the *Show()* method in the base class, which contains no code to write to the screen.

Because *Show()* has no code in it, you really do not need it except as a place holder. You can declare a method as *abstract* simply by adding the *abstract* keyword. You would end the declaration with a semicolon and remove the open and closing braces that form the body of the method:

```
abstract public virtual void Show ();
```

If you forget to override *Show()* in your derived class, the compiler will issue an error and tell you that you must provide an override method.

When you declare *Show()* in this way, it becomes what programmers call a *pure virtual function*. It has a declaration but no body.

WHAT YOU MUST KNOW

In this lesson, you have learned about the basic concepts of object-oriented programming—encapsulation, inheritance and polymorphism. You learned how to use a class definition as a *base* class to provide the foundation for another class, the *derived* class. You learned how to declare *abstract* classes and functions and how to override methods in the derived class using the principle of polymorphism. Finally, you learned how to write a list of enumerated values to provide aliases to constant values. In Lesson 9, "Getting Started With C# Operators," you will learn how to use the various operator symbols in C# to build compound *expressions*—expressions that are more complex than a simple assignment statement. Before you continue with Lesson 9, however, make sure you have learned the following key concepts:

- ✗ Objects may contain properties and methods as well as field variables. The properties and methods provide access to class members so that you may set, retrieve and manipulate the values.

- ✗ C# divides variables into two arbitrary types: *value*-types and *reference*-types. Ordinary fields are *value*-types. Objects such as an instance of a class are *reference*-types.

- ✗ Enumerated lists provide convenient aliases for commonly used constants. You build an enumerated list using the *enum* keyword. Enumerated data types provide no memory storage and may be declared outside of a class definition.

- ✗ Encapsulation refers to the concept of keeping related information in a single object. C# enforces encapsulation by requiring you to make all methods, properties and field members of a class.

- ✗ You can build on the methods and properties of an existing class by deriving a new class. The new class is the "derived" class and *inherits* the methods, properties and fields of the existing class, which is the "base" class.

x By providing *virtual* functions in your base class, you may override methods and properties to modify the way the class operates—creating a polymorphic object that can change forms.

ou have learned about variables and objects in C# and how to write a simple statement, such as an *assignment statement*, to store a value in a variable. At some point, your program will need to perform *operations* on values and the values your variables contain. To perform operations, you use one of the C# *operator* symbols, such as the plus sign (+) to add two values or the asterisk (*) to multiply two values. C# contains a rich set of operators you can use to manipulate numbers and to perform arithmetic. In this lesson, you will learn about those operators and how to use them. By the time you finish this lesson, you will understand the following key concepts:

- To perform arithmetic and logical operations on values and variables, you use the C# operators.
- C# represents most operators using a symbol you type, such as the + and / symbols to represent addition and division. In some cases, such as *sizeof*, a keyword represents an operator.
- The variables and values you use with an operator to perform an operation are the *operands*.
- Most C# operators are *binary* and require two operands. *Unary* operators require only one operand. C# contains only one *ternary* operator, which requires three operands.
- An expression is a combination of operators and values or variables. Your program *evaluates* an expression to get a *result*. In C#, you must assign the result to a variable.

- Relational operators compare two values and return a boolean result of *true* or *false*. You use boolean values to give your program the ability to make decisions.
- Bitwise operators combine the individual bits in two values in a logical sequence.

Reviewing Operators

Nearly any program you write must perform some sort of *operation* on the values and variables you use in the program. It is a rare program that does not use any of the C# operators. Even the simple task of storing a value in a variable, such as *int iVar = 42*, performs an assignment operation using the *assignment operator*, the equals sign.

An example of an operation might be to add two numbers, such as *16 + 8*. In this case, the operator is the plus sign, and the two values, *16* and *8*, are the *operands*. You may use variables or certain other identifiers as the operands. When you combine two or more values, variables or identifiers with an operator, the result is an *expression*. Your program evaluates the expression and returns a *result*. In C and C++, it usually does not matter what you do with the result. You may choose to ignore the result, as in the following statement:

```
16 + 8;
```

Such a statement is not entirely useless in a language such as C or C++. But when you write programs in C#, the compiler will not let you ignore the result, and the preceding statement would cause the compiler to issue an error. You must assign the result to a variable or use it in some other way. If you do not, the compiler will refuse to compile your program. The following statement would make the preceding statement valid in C#:

```
int iVar = 16 + 8;
```

C# inherits a rich set of operators from the C and C++ languages. In fact, there are more operators than there are symbols on your keyboard (excluding, of course, the letter and number keys), so C# combines symbols to represent some operations. For example, you have seen the *assignment* operator in use, but in C# there are 10 other assignment operators, each represented by combining the equals sign with another symbol.

Operators fall in three broad categories. The *unary* operators require only one operand. Some programmers call these operators *monadic* operators. An example is the increment operator (++) that instructs the program to add *1* to the value of the operand.

Most operators require two operands. These are *binary* operators, and some programmers refer to them as *diadic* operators. An example is the addition operator, which causes the program to add two values such as in *16 + 8*.

Whether an operator is unary or binary is only occasionally important, but there are times when you should be aware of the difference. The plus and minus operators, for example, may be unary or binary depending upon how you use the operators. When the type of an operator is important, this book will refer to operators as unary or binary. You may see "monadic" and "diadic" in other books, so you should be aware that these words refer to unary and binary operators, respectively.

Most languages do not have operators that require more than two operands, but C# contains one *ternary* operator, which requires three operands. This operator is the *conditional expression* operator, and you will learn about it in Lesson 20, "Making Decisions Within a C# Program."

Understanding Unary Operators

The *unary* operators perform their operations on a single operand. The C# unary operator set includes symbols to increment (add one to) or to decrement (subtract 1 from) the value of a variable, and those to manipulate the individual bits in a value.

Table 9.1 summarizes the C# unary operators.

Operator	Operation	Result
~	1's complement	Reverses the sense of the bits in a variable. The ones are changed to zeros and the zeros are changed to ones.
!	Negation	Returns *true* if a value is *false* or *false* if a value is *true*.
-	Unary Minus	Returns the negative sense of a value or variable.
+	Unary Plus	Returns the positive sense of a value or variable.
++	Increment	Adds one to the value of a variable
--	Decrement	Subtracts one from the value of a variable
new	Allocate	Reserves a memory location where a reference-type value may be stored.
sizeof	Size of	Returns the number of bytes required to store the value of a variable

Table 9.1 C# unary operators.

The 1's complement operator reverses the value of the individual bits in a number. Early on, you learned that the computer represents values as a sequence of bits or *binary digits*. A bit is either a 1 or a 0. Using the 1's complement operator causes the computer to change all of the 1's in a value to 0's, and all of the 0's to 1's. For example, when you declare a *short* integer variable and assign it the value 15982 as in the following statement, the computer would store the value as a sequence of sixteen 1's and 0's:

```
short sVal = 15982;
```

Internally, the computer would represent the number as 0011111001101110. Occasionally, you may read or hear programmers use the word "not" when they encounter the 1's complement operator. This is jargon, and you should not confuse it with the *negation* operator. Some programmers and texts also use the word "not" for the negation operator.

The following program, *OnesComp.cs*, contains C# concepts you may not have learned yet. The intent is to show you that the 1's complement operator actually reverses the sense of the individual bits in a value:

```
namespace nsOnesComp
{
    using System;
```

```
class clsMain
{
    static void Main ()
    {
        clsMain main = new clsMain();
        short iVar1 = 15982;
        short iVar2 = ~15982;
        Console.WriteLine ("iVar1 = {0}", iVar1);
        Console.WriteLine ("iVar2 = {0}\r\n", iVar2);
        Console.Write ("In binary, iVar1 = ");
        main.ShowBits(iVar1);
        Console.Write ("In binary, iVar2 = ");
        main.ShowBits(iVar2);
    }
    public unsafe void ShowBits (short iVar)
    {
        int Bit = sizeof (short) * 8 - 1;
        for (int i = 0; i < (sizeof (short) * 8); ++i)
        {
            int iBit = (iVar >> Bit - i) & 1;
            Console.Write (iBit);
        }
        Console.Write ("\r\n");
    }
}
```

Be sure to compile this program using the */unsafe* switch. When you compile and run this program, you will see the following output:

```
iVar1 = 15982
iVar2 = -15983

In binary, iVar1 = 0011111001101110
In binary, iVar2 = 1100000110010001
```

A computer uses the highest bit (the first one printed out above) to represent a negative number. If the bit is 0, the value is positive. By toggling this *sign bit*, the value became negative. Notice, however, that the 1's complement is not exactly equal to the negative of the original value. You will see why shortly when you cover the unary minus operator, which programmers often call the 2's complement operator.

The *negation* operator, the "!" symbol, has limited use in C# compared to its power and flexibility in C and C++. In C#, you may use this symbol only on a boolean expression to reverse the sense of the result of the expression. The negation operator returns *false* if the value of the boolean expression is *true*, and *true* if the boolean expression is *false*. If you have studied C or C++, you may have used this operator to test for a non-zero value, as in the following snippet:

```
int iVar = [some value]
if (!iVar)
{
}
```

In C#, this snippet would cause the compiler to issue an error because *iVar* is not a boolean value. As you will see when you study the relational operators, this limitation severely restricts the usefulness of the negation operator.

The *unary minus* operator causes your program to take the negative of the value or variable that follows the symbol. Negating a number using the minus sign is the same concept that you learned in arithmetic—writing –4 means "take the negative of the value of 4."

Programmers sometimes call the *unary minus* operator the 2's complement operator because the 2's complement of a value is the same as its negative value. To find the 2's complement, use the 1's complement operator and add 1 to the result, as shown in the following short program, *TwosComp.cs*:

```
namespace nsTwos
{
    using System;
    class clsTwosComp
    {
```

```
        static void Main()
        (
            int Pos = 15982;
            int Neg = -Pos;
// Get the 1's complement of Pos
            int OnesComp = ~Pos
// Show the values so far
            Console.WriteLine ("Positive value = {0}", Pos);
            Console.WriteLine ("Negative value = {0}", Neg);
            Console.WriteLine ("Ones Complement = {0}",
                                OnesComp);
// Now add 1 to the 1's complement to get the 2's complement
            int TwosComp = OnesComp + 1;
// Show that the 2's complement is the negative of Pos
            Console.WriteLine ("Positive value = {0}", Pos);
            Console.WriteLine ("Twos Complement = {0}",
                                TwosComp);

        }
    }
}
```

When you compile and run *TwosComp.cs*, you will see the following output. Try changing the value to assure yourself that the 2's complement of a number is the same as its negative value.

```
Positive value = 15982
Negative value = -15982
Ones Complement = -15983
Positive value = 15982
Twos Complement = -15982
```

This is the same way the processor chip in your computer gets the negative value of a number. The internal circuitry first gets the 1's complement of a number, then adds 1 to that value to get the 2's complement, or negative value.

The *unary plus* operator essentially does nothing. Its position in the operator table simply provides balance with the unary minus, and allows you to write a value or variable with a plus or minus symbol in front of it. The following two statements are equivalent, and both assign the same value to the variable:

```
int Pos1 = 15982
int Pos2 = +15982
```

You learned how to use the *new* and *sizeof* operators in Lesson 6, "Using Variables to Store Information in C# Programs." However, Lesson 6 did not cover one important aspect of the *sizeof* operation. You may use the *sizeof* operator only in an *unsafe* context. You will learn how to use *unsafe* code later in this lesson.

The *increment* and *decrement* operators provide convenient shorthand for a couple of common statements in programming. You will learn how to use these operators in Lesson 17, "Using Increment and Decrement Operators."

Understanding Binary Operators

The most common operator type in C# is the binary type. Binary operators require two operands. This group of operators includes those you commonly use for arithmetic operations and those that you use to test the relationship between two values.

Table 9.2 summarizes the binary operators.

Operator	Operation	Result
+, -, *, /, %	Arithmetic	The result of the arithmetic operation.
<<, >>	Shift	Shifts the bits (the ones and zeros) in the left operand the number of times specified by the right operand.
<. >, <=, >=, ==, !=, is	Relational	*true* or *false*.
&, |, ^	Bitwise	Combines the bits in the left operand with the bits in the right operand.
&&, ||	Logical	*true* or *false*.
=	Assignment	Sets the value of the left operand to the value of the right operand.

Table 9.2 The C# binary operators.

You might not be familiar with all of these symbols, but you will learn and use the operators throughout the course of this book.

To use a binary operator, you write the value or identifier—the first operand—on the left side of the operand. Then, you must have a value or variable—the second operand—that the operator will apply to the first value. You write this second value to the right side of the operator. In the expression *3 + 4*, the *3* is the starting value and the *4* is the value the operator will apply to the first value. In this case, it will add *4* to *3*.

This order is important. When you write an expression this way, programmers say that the operator *associates* the values from left to right. That is, the program always evaluates the value on the left first, then the value on the right. The program then applies the value on the left to the value on the right using the binary operator.

For a simple expression such as *3 * 4*, this may sound trivial, but the value on each side of the operator might be another expression containing other operators. For example, instead of *3 * 4*, you might write the following expression:

```
int result = (24 / 8) * (64 / 16);
```

Your program first will evaluate *24 / 8*, then *64 / 16* before applying the multiplication (*) operator. The parentheses group values and operators into a single expression; you will learn about expressions in Lesson 18, "Writing Expressions in C#."

Using Arithmetic Operators

You will use the arithmetic operators often in your code.

While most people know how to use the +, -, *, and / operators, the "modulo" operator, the "%" sign, may be unfamiliar to some readers. When you divide one *int* value by another, the result is a whole number, an *int* value. The program simply discards any remainder from the division. The modulo operator, however, performs an integer division and returns the *remainder* of the operation as a result. For example, the result of *7 % 2* is *1* because *7* divided by *2* yields *3* and a remainder of *1*. The *modulo* operation deals only with the remainder. You can only use the modulo operator with integer values.

You might use modulo division in a program to calculate the change and what coins to return a customer in a sales transaction, as in the following program, *Change.cs*:

```
namespace nsChange
{
    using System;
    class clsCountChange
    {
        static void Main ()
        {
            int cost = 1550;        // The cost of an item
            int sales_tax = 6;      // Sales tax is 6%
            int amount_paid = 2000; // Amount the buyer paid
            int tax, change, total; // Sales tax, change
                                    // and total bill

            tax = cost * sales_tax / 100;
            total = cost + tax;
            change = amount_paid - total;

            Console.WriteLine ("Item cost: ${0}.{1}",
                    cost / 100, cost % 100);
            Console.WriteLine ("Tax: ${0}.{1}",
                        tax / 100, tax % 100);
            Console.WriteLine ("Total:  ${0}.{1}",
                        total / 100, total % 100);
            Console.WriteLine ("Customer change:  ${0}.{1}",
                        change / 100, change % 100);
//
// Compute the change in dollars and coins
//
            int cash;
            int dollars, quarters, dimes, nickels, cents;

            cash = change;
            dollars = cash / 100;            // How many dollars
```

```
        cash = cash - dollars * 100;  // Subtract dollars
        quarters = cash / 25;         // How many
                                      // quarters
        cash = cash - quarters * 25;  // Subtract
                                      // quarters
        dimes = cash / 10;            // How many dimes
        cash = cash - dimes * 10;     // Subtract dimes
        nickels = cash / 5;           // How many nickels
        cents = cash - nickels * 5;   // Subtract nickels
                                      // to get pennies
//
// Show the change to return to the customer
//
        Console.WriteLine ("\t{0} dollar bills",
                            dollars);
        Console.WriteLine ("\t{0} quarters", quarters);
        Console.WriteLine ("\t{0} dimes", dimes);
        Console.WriteLine ("\t{0} nickels", nickels);
        Console.WriteLine ("\t{0} pennies", cents);
      }
    }
}
```

This program calculates the transaction in cents so that it uses only integer arithmetic. That's because, you cannot use the *modulo* operator on non-integer values such as a *double*. The item costs $15.50, or 1550 cents, and the customer gives the clerk a $20 bill, or 2000 cents. When you compile and run *Change.cs*, you will see the following output:

```
Item cost: $15.50
Tax: $0.93
Total:  $16.43
Customer change:  $3.57
        3 dollar bills
        2 quarters
        0 dimes
        1 nickels
        2 pennies
```

Try using the *Console.Read()* or *Console.ReadLine()* functions from Lesson 4, "Using C# From the Command Line," to modify the program to accept the sales total and the amount paid from the command line.

Understanding Assignment Operators

You have been using the assignment operator, the equals sign, in your programs in this and previous lessons. You used the operator to tell the compiler that you want your program to place a particular value in a particular memory location. In most computer languages, the equals sign is the only assignment operator.

C#, though, has no fewer than 11 assignment operators. The equals sign allows you to do a *simple* assignment, but the other 10 give you the ability to combine an arithmetic operation with the assignment operation. Table 9.3 summarizes the assignment operators.

Operator	Example	Result
=	Var = val	Assigns *val* to *Var*.
+=	Var += val	Adds *val* to the value of *Var* and assigns the result to *Var*.
-=	Var -= val	Subtracts *val* from the value of *Var* and assigns the result to *Var*.
*=	Var *= val	Multiplies the value of *Var* by *val* and assigns the result to *Var*.
/=	Var /= val	Divides the value of *Var* by *val* and assigns the result to *Var*.
%=	Var %= val	Modulo divides the value of *Var* by *val* and assigns the remainder to *Var*.
<<=	Var <<= val	Shifts the value of *Var* to the left *val* times and assigns the result to *Var*.
>>=	Var >>= val	Shifts the value of *Var* to the right *val* times and assigns the result to *Var*.
&=	Var &= val	Combines the bits in the value of *Var* and *val* using a bitwise *and* operation and assigns the result to *Var*.
|=	Var |= val	Combines the bits in the value of *Var* and *val* using a bitwise *or* operation and assigns the result to *Var*.
^=	Var ^= val	Combines the bits in the value of *Var* and *val* using a bitwise *xor* operation and assigns the result to *Var*.

Table 9.3. The assignment operators used in C#.

A common operation in programming is to perform an operation on the value of a variable, and then to assign the result to the variable that you used in the operation. Assume you have a variable *Var* that contains a value of 14 and you want to add 28 to it to get a value of 42. In most languages you would have to write *var = var + 28* to set the new value for *var*. C#, however, lets you combine one of the binary operators with the assignment operator. In C#, the following statements both add *28* to the value of *var*:

```
Var = Var + 28;
Var += 28;
```

Obviously, the second method reduces the amount of typing you have to do when you write the statement. The less you have to type, the less chance you have of making a typing mistake, such as misspelling the name of a variable.

The statement *Var += 28;* means "take the value of *Var* and add 28 to it, then assign the result of the addition to *Var*." That really is what the first statement, *var = var + 28;* means. The difference is that you do not have to type the variable's name twice.

You may write any of the binary operators except the logical operators (*&&* and *||*) and the assignment operator itself in this shorthand method. Always write the name of the variable first, then the operator, next the assignment operator (the equals sign) and finally the value you want to apply to the variable. The following statements show some examples:

```
var -= 28;      // Subtract 28 from var
var /= 28;      // Divide var by 28
var *= 28;      // Multiply var by 28
var <<= 4;      // Shift the bits in var to the left 4 times
```

The following program, *Assign.cs*, shows how to use the assignment operators and their effects on the value of a variable:

```
namespace nsAssign
{
    using System;
    class clsAssign
```

```
    {
        static void Main ()
        {
            int Var = 2;

            Console.WriteLine ("Starting value for " +
                                "Var = {0}", Var);
//
//  Add 1 to Var
            Var += 1;
            Console.WriteLine ("After Var += 1, " +
                                "Var = {0}", Var);
//
//  Multiply Var by 2
            Var *= 2;
            Console.WriteLine ("After Var *= 2, " +
                                "Var = {0}", Var);
//
//  Divide Var by 2
            Var /= 2;
            Console.WriteLine ("After Var /= 2, " +
                                "Var = {0}", Var);
//
//  Shift the bits in Var three spaces to the left
            Var <<= 3;
            Console.WriteLine ("After Var <<= 3, " +
                                " Var = {0} ", Var);
//
//  Shift the bits in Var three spaces to the right
            Var >>= 3;
            Console.WriteLine ("After Var >>= 3, " +
                                "Var = {0}", Var);
        }
    }
}
```

The program sets the initial value of *Var* to 2, then adds *1* to it using the +=
assignment operator. The result, *3*, is assigned to the value of *Var*. Next, the program uses the *=* and */=* operators and displays the result. You can see, for
example, that writing *Var *=3;* is the same as writing *Var = Var * 3;*.

Notice that some of the operators are made from two operator symbols written together. You *must not* put a space between these two symbols. If you do, the C# compiler will interpret them as two separate operations and issue an error.

Using Relational Operators to Compare Two Values

In C#, like most programming languages, programs make decisions based upon the results of a logical operation using the *relational* operators. These operators compare the relative value of two quantities—either number values or the values stored in variables. Table 9.4 summarizes the relational operators used in C#.

Operator	Operation	Example	Result
<	Less Than	Var1 < Var2	*true* if first operand is less than the second. Otherwise *false*
>	Greater Than	Var1 > Var2	*true* if the first operand is greater than the second. Otherwise *false*
<=	Less Than or Equal	Var1 <= Var2	*true* if the first operand is less than or equal to the second. Otherwise *false*
>=	Greater Than or Equal	Var1 >= Var2	*true* if the first operand is greater than or equal to the second. Otherwise *false*
==	Equal To	Var1 == Var2	*true* if the first operand is equal to the second. Otherwise *false*
!=	Not Equal To	Var1 != Var2	*true* if the first operand is not equal to the second. Otherwise *false*

Table 9.4 The C# relational operators

The result of a relation operation yields no numerical information about the operands themselves. A relational operation always returns a boolean value, either *true* or *false*.

The following command line program, *Compare.cp,* shows the results of using the relational operators using variables equal to *2* and *7* as operands:

```
namespace nsCompare
{
    using System;
    class clsCompare
    {
        static void Main ()
```

```
{
        int var1 = 2;
        int var2 = 7;
        Console.WriteLine ("{0} is less than {1} = {2}",
                            var1, var2, var1 < var2);
        Console.WriteLine ("{0} is greater than " +
                            "{1} = {2}",
                            var1, var2, var1 > var2);
        Console.WriteLine ("{0} is less than or " +
                            "equal to {1} = {2}",
                            var1, var2, var1 <= var2);
        Console.WriteLine ("{0} is greater than or " +
                            "equal to {1} = {2}",
                            var1, var2, var1 >= var2);
        Console.WriteLine ("{0} is equal to {1} = {2}",
                            var1, var2, var1 == var2);
        Console.WriteLine ("{0} is not equal to " +
                            "{1} = {2}",
                            var1, var2, var1 != var2);
    }
  }
}
```

When you compile and run *Compare.cs*, you should see the following output:

```
2 is less than 7 = True
2 is greater than 7 = False
2 is less than or equal to 7 = True
2 is greater than or equal to 7 = False
2 is equal to 7 = False
2 is not equal to 7 = True
```

The "==" (is equal to) and "!=" (is not equal to) operators also are called the *equality* operators because they return *true* or *false* depending upon whether the two operands are equal to each other.

Notice that the "is equal to" (==) operator is *two* equals signs written together. A common error even among programmers with some experience is to write the operator with a single equals sign when writing a *conditional* statement (you will learn about conditional statements in Lesson 20, "Making Decisions Within a C# Program"). In C and C++, the compiler *will not* issue or display an error if you do this. In C#, however, the expression in a conditional statement *must* evaluate to a boolean value, and the compiler will issue an error and refuse to compile your program if you make this sort of error.

As with the assignment operators, some of the relational operators are made from two symbols. You cannot have any spaces between the two symbols.

Logical Operators: Testing for True and False Conditions

The logical operators test the results of two values or expressions. The operators test the relationship between two boolean values and return a boolean value to represent the result.

There are only two logical operators. The logical AND operator returns *true* only if both expressions are *true*. You write the logical AND operator with two ampersands (&&) between the values or expressions. The logical OR operator returns *true* if *either* of the two values or expressions is *true*. You write the logical OR operator with two vertical bars (||) between the two operands. There cannot be any spaces between the ampersands or vertical bars.

In C#, the logical operators test the operands for *true* or *false* values, and return boolean *true* or *false* depending upon the result of the expression. Table 9.5 is a truth table showing the result of the logical operators.

Expr1	Expr2	Expr1 && Expr2	Expr1 || Expr2
true	true	true	true
true	false	false	true
false	true	false	true
false	false	false	false

Table 9.5 Truth table for logical operators.

As with the negation operator, the constraints of the C# language specification have emasculated the logical operators when compared to the power and usefulness they enjoy in C and C++. Both operands must be boolean values; you cannot use them to test simply whether two values are zero or non-zero. Thus, the statements in which you use them sometimes might appear convoluted. The following program, *Logical.cs*, shows samples of the logical operators:

```
namespace nsLogical
{
    using System;
    class clsLogical
    {
        static void Main ()
        {
            int var1 = 0;
            int var2 = 2;
            Console.WriteLine ("{0} AND {1} = {2}",
                    var1 != 0, var2 != 0,
                    (var1 != 0) && (var2 != 0));
            Console.WriteLine ("{0} OR {1} = {2}",
                    var1 != 0, var2 != 0,
                    var1 != 0 || var2 != 0);
//
// Set new values for var1 and var2
//
            var1 = 4096;
            var2 = -8192;
            Console.WriteLine ("{0} AND {1} = {2}",
                    var1 != 0, var2 != 0,
                    var1 != 0 && var2 != 0);
            Console.WriteLine ("{0} OR {1} = {2}",
                    var1 != 0, var2 != 0,
                    var1 != 0 || var2 != 0);
        }
    }
}
```

When you compile and run *Logical.cpp*, you will see the following output:

```
False AND True = False
False OR True = True
True AND True = True
True OR True = True
```

Understanding Bitwise Operators

The *bitwise* operators combine the bits of one operand with those of a second operand. The operators work on corresponding bits in the two operands. For example, the lowest order bit, bit 0, of the first operand is combined with the lowest order bit of the second operand using the logical operation. Then bit 1 of the first operand is combined with bit 1 of the second operand. The operation is complete when all the bits have been combined.

Table 9.6 summarizes the three bitwise operators, and Table 9.7 is a *truth table* for the bitwise operations.

Operator	Operation	Result
&	AND	1 if both bits are 1, otherwise 0
|	OR	1 if either bit is 1, otherwise 0
^	XOR	1 if both bits are *different*, otherwise 0

Table 9.6 The C# bitwise operators.

Programmers sometimes call the XOR operators the "exclusive OR" operator. As you can see from the truth table, the result is 1 only if one and only one exclusively is 1. If both bits are 1 or both bits are 0, the result is 0.

Bit1	Bit2	Bit1 & Bit2	Bit1 | Bit2	Bit1 ^ Bit2
1	1	1	1	0
1	0	0	1	1
0	1	0	1	1
0	0	0	0	0

Table 9.7 The truth table for the C# bitwise operators.

The XOR operator has some special properties. Programmers sometimes call it the "toggle operator." If you XOR a bit with 1, the result is always the *opposite* of the original bit. For example, 1 ^ 1 is 0, and 0 ^ 1 is 1.

The following program, *Bitwise.cpp*, gives examples of the bitwise operators. The values and the results are shown in *binary* form, using the ones and zeros as they are stored in the memory location. The program contains a function marked unsafe, so you must compile the program using the */unsafe* switch.

```cpp
namespace nsBitwise
{
    using System;
    class clsBitwise
    {
        static void Main ()
        {
            clsBitwise Bwise = new clsBitwise();
            int var1 = 7324;
            int var2 = 1693;

            Console.Write ("        var1 = ");
            Bwise.ShowBits (var1);
            Console.Write ("        var2 = ");
            Bwise.ShowBits (var2);
            Console.Write ("var1 & var2 = ");
            Bwise.ShowBits (var1 & var2);
            Console.WriteLine ();

            Console.Write ("        var1 = ");
            Bwise.ShowBits (var1);
            Console.Write ("        var2 = ");
            Bwise.ShowBits (var2);
            Console.Write ("var1 | var2 = ");
            Bwise.ShowBits (var1 | var2);
            Console.WriteLine ();

            Console.Write ("        var1 = ");
            Bwise.ShowBits (var1);
            Console.Write ("        var2 = ");
            Bwise.ShowBits (var2);
```

```
            Console.Write ("var1 ^ var2 = ");
            Bwise.ShowBits (var1 ^ var2);
        }

        public unsafe void ShowBits (int iVar)
        {
            int Bit = sizeof (int) * 8 - 1;
            for (int i = 0; i < (sizeof (int) * 8); ++i)
            {
                int iBit = (iVar >> Bit - i) & 1;
                Console.Write (iBit);
            }
            Console.Write ("\r\n");
        }
    }
}
```

When you compile and run *Bitwise.cp*, you should see the following output. Compare the bits of the original variables with the bits of the result of the operation using the Table 9.7.

```
         var1 = 00000000000000000001110010011100
         var2 = 00000000000000000000011010011101
 var1 & var2 = 00000000000000000000010010011100

         var1 = 00000000000000000001110010011100
         var2 = 00000000000000000000011010011101
 var1 | var2 = 00000000000000000001111010011101

         var1 = 00000000000000000001110010011100
         var2 = 00000000000000000000011010011101
 var1 ^ var2 = 00000000000000000001101000000001
```

Try changing the values of *var1* and *var2* and recheck the results. Assign a negative value to one of the variables and compare the results again.

Using Unsafe Code

In this lesson's previous examples, you have seen the keyword *unsafe* in combination with a function that uses the *sizeof* operator. To understand the *unsafe* keyword, you need to know a little about *pointers*. Normally, you will not use pointers in C# code, so this book will not spend a lot of time discussing them. For a deeper understanding of pointers, you should consult a C++ programming book such as *Starting with Visual C++*.

You have learned that a variable is a memory location where your program stores a value. Your program creates value-type variables on the stack and reference-type variables in the heap using the *new* operator. Sometimes in C and C++ it is more convenient to refer to the variables by their location in memory rather than by their identifier names. A pointer is just that—a value that is the address in memory where your program stores a variable's value.

A variable that holds the address of another variable is a *pointer variable*. You declare such a variable using an asterisk in front of the identifier for the variables. The following declaration says that *pVar* is a variable that can hold the address of another variable of type *int*:

```
int *pVar;
```

To get the memory location of a variable, you use the *address* operator, an ampersand, before the variable identifier. You can assign the address to a pointer variable. The following snippet declares a variable, *iVar*, and assigns the variable the value *42*. Then it declares a second variable, the pointer variable *pVar*, and assigns it the address of *iVar*. Now *pVar* contains the actual memory location where your program stores the value of *iVar*.

```
int iVar = 42;
int *pVar = &iVar;
```

To retrieve a value through a pointer variable, you have to *dereference* the variable using the *indirection operator*, also an asterisk:

```
int iNewVar = *pVar;
```

Pointers are an extremely important concept in C. The only way to access a variable in the heap in C is through a pointer to the variable. The importance of pointers dropped only slightly in C++, which uses *reference variables*. A reference variable is really a pointer variable that acts like an ordinary variable.

Normally in C#, the Common Language Runtime manages objects you create on the heap and takes care of handling the pointers for you. However, the Common Language Runtime does not manage code that you declare as *unsafe* and you may declare and use pointers in *unsafe* code. You also must declare code as *unsafe* when you use the *sizeof* operator, although the MSDN help file and the language specification do not really give a reason for this.

You can declare an entire function as *unsafe* by adding the keyword to the function's declaration before the function's return type. If you use an unsafe function or statement, you must compile your program using the */unsafe* switch. In the following function, you may use pointers and the *sizeof* operator anywhere in the function:

```
public unsafe void ShowBits (int iVar)
{
}
```

You also may declare a *block* of code as *unsafe* by using the keyword followed by the unsafe code within a set of braces. You could have removed the *unsafe* declaration from the *ShowBits()* function in the last section by writing it as follows, using the *sizeof* operator only within an *unsafe* block:

```
public void ShowBits (int iVar)
{
    int Bit;
    int Max;
    unsafe
    {
        Bit = sizeof (int) * 8 - 1;
        Max = sizeof (int) * 8;
    }
    for (int i = 0; i < Max; ++i)
    {
        int iBit = (iVar >> Bit - i) & 1;
        Console.Write (iBit);
    }
    Console.Write ("\r\n");
}
```

Finally, you could declare just a single statement as *unsafe* by using the *unsafe* keyword at the beginning of the statement:

```
int Bit;
unsafe Bit = sizeof (int) * 8 - 1;
```

Before you try to use pointers inside an *unsafe* block, you should consult a C++ text and practice using pointers in C++. If your program requires extensive use of pointers, then the application might be more suited to C++ than to C#.

WHAT YOU MUST KNOW

In this lesson, you learned how to use operators in C#. You learned how to combine operators with values and variables to write expressions. In addition, you learned how to use unary operators, including the increment and decrement operators. You learned how to combine assignment and arithmetic operations using the rich set of assignment operators in C#, and how to manipulate the individual bits in a variable using operators. In Lesson 10, "Understanding

Constructors and Destructors," you will learn how to use operators and expressions to perform statements automatically when you declare a class object. Before you continue with Lesson 10, however, make sure you have learned the following key concepts:

- C# uses operators to perform arithmetic and logical operations on values and variables. You type an operator using one or more keyboard symbols. C# represents a few operators using keywords.

- Operators require one or more operands, values or variables that the operator uses to perform the operation.

- When you group one or more values, variables and operators together, you are writing an expression. Your program evaluates the expression and returns a result.

- In C#, you may not ignore the result of an expression. You must assign the result to a variable or use it in some other way such as in a function call or as part of another expression.

- C# contains a rich set of operators that you may use in assignments, in logical expressions or to manipulate the individual bits of a value.

- The Common Language Runtime does not manage code that you declare as *unsafe*. You may mark an entire function, a block of code or just a single statement as *unsafe*.

UNDERSTANDING CONSTRUCTORS AND DESTRUCTORS

a *class* in C# is an *object*. Objects have special status in object-oriented programming. An object may contain methods, properties, and variables that you may create using a single declaration and manipulate by accessing the class object. Classes also have special functions—*constructors*—that you may use to initialize instances of the class when you first create them. In addition, classes have *destructors*, functions that execute when the Common Language Runtime destroys the object. In this lesson, you will learn how to write constructors and destructors for your classes. You will learn how to initialize *readonly* properties and fields in your class using constructors. By the time you finish this lesson, you will understand the following key concepts:

- You must create instances of a class by using the *new* operator. The Common Language Runtime, in turn, will manage the class instance.

- You may specify parameters to be used by your class when you declare an instance. The Common Language Runtime passes the parameters to a special function, a *constructor*, that executes automatically.

- Your class may contain more than one constructor. The Common Language Runtime will determine which constructor to call when you create an instance of your class. Your class also may contain a *static* constructor through which you may initialize *static* members without creating a class instance.

- The Common Language Runtime manages instances of a class. When your program no longer needs an instance of a class, the Common Language Runtime *garbage collector* destroys the class instance.

- When the Common Language Runtime destroys a class instance, a special function, a *destructor*, executes automatically. You may use the destructor to free system resources that your class instance uses.
- The Common Language Runtime may not destroy unneeded class instances immediately when your program no longer needs them. Depending upon the amount of memory in your computer, the class instance may exist for some time before the Common Language Runtime destroys it.
- You may not call a destructor function directly. If you must delete objects in a class before the Common Language Runtime destroys the object, you can provide a *Dispose()* function.

Using the *new* Operator

In C#, the only method you may use to create an instance of a class is by using the *new* operator. The Common Language Runtime manages objects in C#, and using the *new* operator reserves memory in the heap for the class instance. After the memory is reserved, the Common Language Runtime will begin managing the object. There is no method to create a class instance on the stack.

In its simplest form, you may declare a variable name without creating an instance of the class. To do this, simply use the class name followed by the identifier:

```
clsClassName ClassVar;
```

The preceding declaration only reserves the variable *ClassVar* through which you will access the class members. It *does not* actually create an instance of the class. If you try to access a class member using *ClassVar* at this point, the compiler will issue an error that you are trying to use "unassigned variable ClassVar." You still must use the *new* operator to create the class instance:

```
ClassVar = new clsClassName();
```

You may combine the declaration and assignment in the same statement, as you have done in previous samples in this book:

```
clsClassName ClassVar = new clsClassName();
```

After this statement, you may begin accessing member methods, properties, and fields using the *member* operator, which is simply a period between the object's variable name and the member name. The following declares a class object, then sets the member variable *var* to 0:

```
class clsMain
{
    static void Main()
    {
        clsMain NewClass = new clsMain();
        NewClass.var = 0;
    }
    public int var;
}
```

You should notice the use of the open and close parentheses in the statements that create an instance of a class. You may include any arguments that your class needs in a parameter list within the parentheses. You *must* include the parentheses even if you do not pass any arguments to the class instance.

When you create the class instance, the Common Language Runtime calls a special function that you may use to initialize variables and properties in your class. This special function is the *constructor*, which executes automatically when you create the class instance.

Understanding Constructors

Often when you create a class, you will include member properties and fields that you must initialize at the same time you create the class. The C# compiler will not let you use any member properties or variables until you have given them an initial value.

A constructor is not an ordinary member function, nor do you declare or use constructors like ordinary functions. You may not call a constructor directly, for example. You must let the CLR call the function when you create a class instance.

A constructor may not have a return type. This means that you cannot declare a constructor that returns an error code to indicate whether it succeeded in initializing member variables and objects. This means that you must provide some other method of determining whether the constructor was successful, such as throwing an *exception* on failure.

To create an instance of a class directly, you must give the constructor an access of *public*. You will learn more about *public* and other access keywords in Lesson 11, "Understanding Class Scope and Access Control." You may use other access levels under certain conditions, but your class generally will have at least one *public* constructor.

A constructor function must have the same name as the class. This identifies the function to the compiler as a constructor and suppresses certain error messages that the compiler might issue for ordinary functions.

The following snippet summarizes these properties of a constructor by declaring a class and one *public* constructor:

```
class clsSample
{
    public clsSample()
    {
        //  Statements to initialize fields and properties
    }
}
```

Note: *If you also program in Visual Basic, you probably are aware of the Initialize event, which serves the same purpose as a constructor when you create an instance of a class. Visual Studio 7, however, introduces the concept of constructors and destructors to Visual Basic classes. Consult the "Initialization and Termination of Components" topics in the MSDN help file for more information.*

Using Constructors

You do not have to write a constructor function for a class if you do not need one. If you do not write a constructor, the compiler will assume a default *public* constructor that will permit you to declare an instance of the class without using any arguments.

If you do write a constructor that needs arguments, however, the compiler will not assume a default constructor, and you must pass arguments to the constructor when you create the class instance.

This is easier to visualize in code. The following snippet defines a class that contains a constructor. The constructor requires a single argument, an *int* value. When you create an instance of the class, you must pass an integer value in the statement that contains the *new* operator:

```csharp
using System;
class clsMain
{
    static void Main()
    {
        int parm = 42;
        clsTest NewClass = new clsTest(parm);
        NewClass.ShowVar();
    }
}
class clsTest
{
    public clsTest(int arg)
    {
        m_Var = arg;
    }
    private int m_Var;
    public void ShowVar()
    {
        Console.WriteLine ("m_Var = {0}", m_Var);
    }
}
```

If you now try to create an instance of *clsTest* without passing an argument, the compiler will issue an error that there is no constructor that takes zero arguments. As you will see shortly, a class can have more than one constructor, and you could provide for a case where you want to create a class instance without passing an argument.

If you do not provide an access keyword (the *public* in the preceding example) for the constructor, the access will default to *private* and you will not be able to declare *any* instances of the class. There are times that you may want to use a *private* constructor. For example, if you use a class simply as a container for *static* properties and fields, declaring an instance of the class would only waste memory.

A C# Class May Have static *Constructors*

Unlike C++, a C# class may declare a **static** constructor. Because it is a **static** member, such a constructor has all the restrictions placed on **static** members that you learned about in Lesson 7, "Getting Started with Classes." Constructors declared as **static** may access only **static** properties, fields and methods in the class.

There also are other restrictions on **static** constructors. A **static** constructor may **not** have an access level (**public**, **protected**, or **private**). The parameter list must be empty. In addition, you may never call a **static** constructor, even when you declare an instance of a class.

The Common Language Runtime will determine when to call a **static** constructor. The C# language specification does not guarantee when the Common Language Runtime will call the constructor, only that the call will occur **before** you declare any instances of the class containing the **static** constructor and **before** you access any other **static** member of the class.

A **static** constructor is handy when you use the class only as a container for **static** properties, fields and methods. You may provide a **private** constructor to prevent any instances of the class, yet you may still use the **static** constructor to initialize the **static** members.

The following program, **Static.cs**, defines a class that contains only **static** members, including a **static** constructor. The constructor prints a message when the Common Language Runtime calls it:

```csharp
namespace nsStatic
{
    using System;
    class clsMain
    {
        static void Main ()
        {
            Console.Write ("clsStatic.sVar = {0}",
                            clsStatic.m_sProp);
        }
    }
    class clsStatic
    {
        static clsStatic()
        {
            Console.WriteLine ("Called static constructor");
            m_sProp = 42;
        }
        static public int m_sProp
        {
            get
            {
                return (m_sVar);
            }
            set
            {
                m_sVar = value;
            }
        }
        static private int m_sVar;
    }
}
```

Even though you never create an instance of **clsStatic**, the Common Language Runtime calls the **static** constructor before you access the member property **m_sProp**.

Using Multiple Constructors

A class may have more than one constructor function. There are times when you will want to be able to pass arguments of different data types to a constructor, depending upon how you want to use the class object. You can provide multiple constructors through *overloading*.

You will learn more about overloading in Lesson 33, "Overloading Functions and Operators." For now, you must understand that each constructor that you write must have a unique *signature*. The signature is a combination of the constructor's name and the data types of its parameter list.

Because each constructor's name must be the same as the class name, you must change the parameter list to change the constructor's signature. The parameter lists of the overloaded constructors must differ in their data types or the number of parameters.

If you provide a constructor that needs two *int* parameters, you cannot create another constructor that needs only two *int* parameters, even if you give the parameters different names. You could, though, write another constructor that uses an *int* and a *long*, because the data types of the parameters would be different. Or you could write another constructor that needs three *int* parameters, in which case the number of parameters would be different.

(The requirement that the parameter list for each constructor must be different does not apply to a *static* constructor. In a *static* constructor, the parameter list *must* be empty, and you can declare only a single *static* constructor. You still may declare a non-*static* constructor that has an empty parameter list.)

You should understand that if you provide multiple constructors, one and only one constructor will execute when you create a class instance. (Again, this does not apply to a *static* constructor. If you provide a *static* constructor and a non-*static* constructor, both will execute.)

The following program, *Construc.cs*, declares multiple constructors for a class, then creates an instance of the class using each constructor. Each constructor prints a message to the screen when it executes:

```
namespace nsConstructor
{
    using System;
    class clsMain
    {
        static void Main()
        {
            int var1 = 42;
            int var2 = 43;
            long var3 = 44;
            double var4 = 45;
            clsTest Test1 = new clsTest ();
            clsTest Test2 = new clsTest (var1);
            clsTest Test3 = new clsTest (var1, var2);
            clsTest Test4 = new clsTest (var3);
            clsTest Test5 = new clsTest (var1, var3);
            clsTest Test6 = new clsTest (var4);
        }
    }

    class clsTest
    {
        static clsTest()
        {
            Console.WriteLine ("Called static constructor");
        }
        public clsTest()
        {
            Console.WriteLine ("This constructor called" +
                                " with no parameters");
        }
// Declare a constructor that needs one int
        public clsTest(int var)
        {
            Console.WriteLine ("Constructor with 1 int");
        }
// Declare a constructor that needs two ints
        public clsTest(int var1, int var2)
        {
            Console.WriteLine ("Constructor with 2 ints");
        }
```

```
// Declare a constructor that needs one long
    public clsTest(long var)
    {
        Console.WriteLine ("Constructor with 1 long");
    }
// Declare a constructor that needs one double
    public clsTest(double var)
    {
        Console.WriteLine ("Constructor with 1 double");
    }
// Declare a constructor that needs one int and one long
    public clsTest(int var1, long var2)
    {
        Console.WriteLine ("Constructor with 1 int " +
                            "and 1 long");
    }
    }
}
```

When you compile and run *Construc.*cs, you will see the following output. Notice that no matter the order in which you create the class objects, the *static* constructor always executes first:

```
Called static constructor
This constructor called with no parameters
Constructor with 1 int
Constructor with 2 ints
Constructor with 1 long
Constructor with 1 int and 1 long
Constructor with 1 double
```

Understanding Destructors

In addition to constructors, a C# class may contain a destructor, a function that executes when the Common Language Runtime destroys the class object. The destructor function gives your class a chance to free up any system resources such as graphics objects that it created.

The Common Language Runtime destroys an object when the garbage collection determines that the object no longer can be reached by a program that is in memory and executing. As with the *static* constructor, you never can know when the destructor will execute, and you have no control over the object destruction. If you have a lot of memory in your computer, the Common Language Runtime may delay the destruction process for some time. If your class is using a lot of system resources or you have declared many instances of a class, you could tie up these resources—and the memory used by the class instances—until the garbage collection decides to remove the objects.

Fortunately, there are workarounds for these quirks. First, you need to learn how to write destructors.

In C++, the traditional method—indeed, the *required* method—to write a destructor is to write a function with the same name as the class name, but precede the name with a tilde. If the class name is *clsClassName*, the destructor would be *~clsClassName()*. The destructor cannot have a return type or an access specifier, and it cannot have a parameter list. A class may have one and only one destructor. The following is an example of a destructor:

```
~clsClassName()
{
    //  Statements to provide cleanup code
}
```

C# is not old enough to have traditions, but you may write a destructor in the same way. The more "traditional" method would be to provide a *Finalize()* method. You cannot do both. If you provide a *Finalize()* method, you cannot write a destructor using the tilde-plus-name sequence. The C# compiler seems to alias the destructor name with *Finalize()*, and providing both functions would cause the compiler to issue an error telling you that *Finalize()* is a duplicate function. The following would cause an error:

```
class clsClassName
{
    ~clsClassName()
```

```
    {
    //   Statements to provide cleanup code
    }
    protected override void Finalize()
    {
    //   Statements to provide cleanup code
    }
  }
```

If you use *Finalize()* as the destructor, you must declare it as shown in the preceding snippet: *protected override void Finalize()*. The reason is that *Finalize()* is a *virtual* function in the *Object* class, and you must override it. When you do, you must be careful not to call the *Object* class function. With all of the typing involved, it is easier to provide the tilde and class name function.

Because *Finalize()* must be declared *protected*, you cannot call it from outside the class. You must provide a *public* method that calls the destructor. The following program, *Destruct.cs*, declares a class with a constructor, a destructor and a *public* method named *Dispose()* that calls the destructor:

```
namespace nsDestruct
{
    using System;
    class clsMain
    {
        static void Main()
        {
            clsTest Test = new clsTest ();
// Other program statements here.
            Test.Dispose ();
        }
    }

    class clsTest
    {
// Declare a constructor
        public clsTest()
        {
```

```
        Console.WriteLine ("Constructor called");
    }
// Define a destructor ~clsTest()
    ~clsTest()
    {
        //  Provide cleanup code here
        Console.WriteLine ("Destructor called");
    }
    public void Dispose()
    {
        Finalize();
    }
    }
}
```

When you must release the resources your class uses, call the *Dispose()* method, which in turn will call the protected *Finalize()* method.

To keep track of any changes in the way C# uses destructors, be sure to read the current "Initialization and Termination of Components" topic in your version of the MSDN help file.

WHAT YOU MUST KNOW

In this lesson, you learned that the Common Language Runtime manages objects when you create them using the *new* operator. You learned how to write constructors, functions that execute automatically when you create an instance of a class. You also learned how to provide your class with multiple constructors, including a *static* constructor, and how to pass arguments when you create a class instance. Finally, you learned about constructors and how the Common Language Runtime destroys objects when the object no longer can be accessed by your code. In Lesson 11, you will learn how to control access to members of a class by using the access keywords and about the *scope* of objects in a class. Before you continue with Lesson 11, however, make sure you have learned the following key concepts:

- The *new* operator creates reference-type objects on the heap. These objects are managed by the Common Language Runtime code.

- A constructor function runs automatically when you create an instance of a class. If you do not provide a constructor function, the compiler will assume a default, empty constructor.

- A class may contain more than one constructor function, but each must have a unique *signature*. You may use additional constructors to pass arguments to a class instance when you create the instance.

- A class may contain a *static* constructor function, but the Common Language Runtime will determine when the function executes.

- When your program no longer can access an object, the Common Language Runtime garbage collector may destroy the object.

- Your may provide your class with a *destructor* function to provide cleanup code when the Common Language Runtime destroys a class object.

UNDERSTANDING CLASS SCOPE AND ACCESS CONTROL

i n most of the sample code so far, you have declared all the members of your classes—methods, fields, and properties—as *public*. This is not always desirable, however, and there will be many times when you will want to protect a class member by giving it another access level. As you have learned, if you do not specify an access, it defaults to *private*, which is not always desirable, either. During your early efforts at programming in C#, you may find yourself declaring most of your class members *public* to simplify your coding. However, as you begin making class definitions available to other programs using *assemblies*, you will want to use other access keywords to limit access to class members. Determining the exact access level to give a class member is a skill that you will acquire in time, and sometimes the process perplexes even experienced programmers. In this lesson, you will begin to learn about the C# access keywords and how the keywords affect how you read and modify fields. By the time you finish this lesson, you will understand the following key concepts:

- The access keywords limit the statements that may retrieve and set the member fields of a class, and control what functions may execute other functions that are members of a class.

- C# defines five different access levels: *private*, *protected*, *public*, *internal*, and *protected internal*.

- If you do not specify an access level for a member field, property, or method, the access defaults to *private* and only other member functions within that class may access the member.

- Commonly, programmers protect member fields and properties by setting the access levels to *protected* or *private*, then write *public* methods through which functions in other classes may modify the values.

- The range of code that may access a field, property, or method is its *scope*. The members of a class are in scope only while an instance of a class exists and is accessible.

Reviewing Class Objects

In the last few lessons, you have learned about the C# class and how to define and implement a class. You learned how to use constructors and how to pass arguments to the constructors when you declare an instance of a class.

When you define a class, you are defining a new data type—a *user-defined* data type. Because a class becomes a data type, you may declare instances of a class the same as you would any other reference data type. To declare an instance of a class, you use the class name as the data type followed by the identifier you want to use as the variable name:

```
clsClassName  ClassVar;
```

A declaration simply reserves the identifier—the variable name—and notifies the compiler that you intend to create a class instance using the identifier. To actually create the class instance, you must use the *new* operator:

```
ClassVar = new clsClassName();
```

You may combine the declaration with the statement that creates the class object:

```
clsClassName ClassVar = new clsClassName();
```

When you create a class object, you create copies of all the fields and properties that are members of the class, and you give the member methods (functions defined within the class) access to those copies.

Often, however, you will want to limit the access to class members to protect them against accidental changes, or to prevent unwanted changes. For example, you might have a class that describes a savings account and contains a field that holds the balance. You probably would want to make sure that your program does not accidentally modify it with an improper operation, such as a negative deposit. You could restrict the access to the balance field and provide access through a method that would prevent such changes.

To restrict access to the class members, C# uses *access* keywords. You have seen examples of access keywords in previous lessons. Table 11.1 summarizes the C# access keywords.

Keyword	Access
public	Any function in any class may access the variable or function.
protected	Only functions in the containing class or *derived* classes may access the variable or function.
private	Only functions in the containing class may access the variable or function.
internal	Only functions in classes within the current module may access the function, or variable.
protected internal	Only functions in the containing class and derived classes in the current project may access the function or variable.

Table 11.1 C# access keywords and how they limit the use of fields, properties, and methods.

Determining which access level to apply sometimes may seem like a dark art. Sometimes it might be obvious, and at other times you might not be sure. Only experience will teach you, and even experienced programmers find themselves adjusting the access of class members.

When you declare a member *public*, there are no restrictions on access to the member. You will find that many of your member methods and properties will need a *public* access so that you may call the method or set the property from another class.

You may restrict public access to a class member to code in the current *module*—the program file that contains the class—by using the *internal* keyword instead of *public*. Using *internal*, functions in other code modules will not be able to access the member. The *internal* keyword will become more obvious when you get to Lesson 13, "Using References and Assemblies" and begin using a code module as an *assembly*.

If you plan to derive one or more classes from a *base* class, and your derived classes will need access to a member of the base class, you should declare the member *protected*. Using *protected internal* restricts the access to class members and to members of derived classes within the code *module*, or the program file in which the class is located. If, for example, you incorporate your module into another project using a *reference*, any classes you derive from the base class in another module may not access the class member. You will learn more about deriving one class from another in Lesson 12, "Deriving a New Class from an Existing Class."

If you do not use an access keyword when you declare a class member, the access will default to *private*. This means that only functions that themselves are members of the class may access the member. The *private* keyword is the highest level of access restriction. Often, you must protect your data fields by declaring them *private*, and then provide *protected* or *public* methods or properties to modify them from functions outside the class.

Establishing Function Access

The access keywords are a primary part of the C# concept of *encapsulation*. A class contains its own member fields, properties and methods, and is able to limit what statements may access its members.

By using access keywords, you may set *rules* on how outside functions may access members of a class to assure that external functions do not assign invalid values to member fields and to assure that variables that are dependent upon one another are up-to-date.

If you want the code in any class in your project to be able to execute a function in a class, you would declare the function *public*. You will often use *public* functions to change values of *protected* or *private* fields in a class. Using class functions to change class variables lets you to provide error checking in your code before the program actually changes the value.

The following program, *Account.cs*, creates a class, *clsAccount*, that contains a *private* field to hold the account balance. You then adjust the balance through *public* functions that first check whether the transaction amount is less than 0. If the transaction amount is negative, the functions print error messages and do not adjust the balance:

```csharp
namespace nsAccounts
{
    using System;
    class clsMain
    {
        static void Main ()
        {
            clsAccount savings = new clsAccount();
            while (true)
            {
                Console.WriteLine ("\nYour savings " +
                                    "balance is now {0:c}\n",
                                    savings.GetBalance());
                Console.WriteLine ("Select one of the " +
                                    "following:");
                Console.WriteLine ("\tFor deposit enter" +
                                    "   1:");
                Console.WriteLine ("\tFor withdrawal enter" +
                                    " 2:");
                Console.Write ("Enter Your selection " +
                                    "(0 to exit): ");
                string strSelection = Console.ReadLine ();
                if (strSelection.Length == 0)
                    continue;
```

```csharp
            int iSel = Convert.ToInt32(strSelection);
            string strAmount;
            if (iSel == 0)
            {
                Console.WriteLine ("\nGoodbye!");
                return;
            }
            else if (iSel == 1)
            {
                Console.Write("Enter deposit amount: ");
                strAmount = Console.ReadLine ();
                decimal Deposit = Convert.ToDecimal(strAmount);
                savings.AddDeposit (Deposit);
            }
            else if (iSel == 2)
            {
                Console.Write("Enter withdrawal: ");
                strAmount = Console.ReadLine ();
                decimal Withdraw =
                        Convert.ToDecimal(strAmount);
                savings.SubWithdrawal (Withdraw);
            }
            else
            {
                Console.WriteLine("Invalid selection\n");
            }
        }
    }
}


    class clsAccount
    {
        private decimal Balance = 0;
// Do not allow negative withdrawals or deposits
        public void AddDeposit(decimal Deposit)
        {
            if (Deposit < 0)
            {
                Console.WriteLine ("Invalid Amount");
                return;
            }
            Balance += Deposit;
```

```
        }
        public void SubWithdrawal(decimal Withdraw)
        {
            if (Withdraw < 0)
            {
                Console.WriteLine ("Invalid Amount");
                return;
            }
            Balance -= Withdraw;
        }
// Return the private balance to the caller
        public decimal GetBalance()
        {
            return (Balance);
        }
    }
}
```

You will learn about *loop* and *conditional* statements throughout the course of this book, but for now you should be aware that the *while(true)* statement in the example makes the code repeat indefinitely, and the successive *if* statements help you to determine the value you entered from the keyboard. For this example, you should concentrate on the access keywords.

If you want to keep functions outside the class from executing a function, you would declare it *protected* or *private*. An access of *protected* would permit the code in *derived* classes to access the function as well. You will learn more about derived classes and how to use *protected* functions in Lesson 12, "Deriving a New Class from an Existing Class." You will use this same class, *clsAccount*, to derive new classes for savings and checking accounts. Giving the function an access of *private* would prevent even the code in derived classes from accessing the function. Only functions within the class may execute other functions that you declare *private*.

In the *Account.cs* example, the functions to adjust the balance keep you from adjusting the balance with a negative value. But a deposit is a positive adjustment, and a withdrawal is a negative adjustment, so you could pass the proper value to a *private* function. Change *clsAccount* as follows:

```
    class clsAccount
    {
        private decimal Balance = 0;
// Do not allow negative withdrawals or deposits
        public void AddDeposit(decimal Deposit)
        {
            if (Deposit < 0)
            {
                Console.WriteLine ("Invalid Amount");
                return;
            }
            AdjustBalance (Deposit());
        }
        public void SubWithdrawal(decimal Withdraw)
        {
            if (Withdraw < 0)
            {
                Console.WriteLine ("Invalid Amount");
                return;
            }
            AdjustBalance (-Withdraw);
        }
        public decimal GetBalance()
        {
            return (Balance);
        }
        private void AdjustBalance (decimal Amount)
        {
            Balance += Amount;
        }
    }
}
```

By the time your code executes *AdjustBalance()*, you know that the dollar amount is correct, and you can use one of the advanced C# assignment operators to adjust the balance.

Accessing Data Members

In a fully developed project involving the *clsAccount* class, the balance would be a very important member. You could never allow it to become negative, and its value would determine whether you could withdraw a given amount (although there are times we wish we could withdraw any amount).

Giving a field an access of *public* allows unlimited access to the field. Had you declared the *Balance* field *public*, it would have no protection. Any statement in any function in your project could change the value, and even make it negative. For this reason, it is common practice to keep important data fields *private* or *protected* and allow only member functions to modify the values.

As with functions, an access of *protected* allows the code in derived classes to change the value of a field. Sometimes this is necessary, but you may restrict access to a field by giving it an access of *private*, which would allow only functions in the same class to access and modify the value.

In the *clsAccount* example, you keep the *Balance* field *private*, but it still is possible for you to enter a withdrawal that would make *Balance* negative. To prevent this, you could modify the *SubWithdrawal()* function:

```csharp
public void SubWithdrawal(decimal Withdraw)
{
    if (Withdraw < 0)
    {
        Console.WriteLine ("Invalid Amount");
        return;
    }
// Do not permit the balance to go negative
    if ((Balance - Withdraw) < 0)
    {
        Console.WriteLine ("You do not have " +
                "enough in your account to " +
                "withdraw {0:c}\n", Withdraw);
        return;
    }
    AdjustBalance (-Withdraw);
}
```

By making *Balance* a *private* field and restricting access to member functions, there is no way your code will let you create a negative balance by entering values from the keyboard. Of course, you could defeat this protection by writing a member function in *clsAccount* that would override the protection.

Understanding Class Scope

Before leaving this lesson, you must understand the lifetime of the functions, fields, and properties in a class, and where in your code you may access the members.

The extent of the code from which you may access the members of a class is the *scope* of the variable you used to store the class instance. You must have a basic understanding of scope as you experiment with your code.

Determining the scope of a variable sometimes can be tricky in C and C++, but in C# it is relatively simple. The scope is the code within the set of braces that encloses the variable declaration. Programmers call this *block* scope because the only code that can access a variable is code that is within the block delimited by the enclosing braces.

The reason for this is that the variable that contains the class does not exist until you declare an instance of the class. When your code exits the block, the variable goes *out of scope* and you cannot access the class instance. Unlike C and C++, the class instance still is in memory and will continue to exist until the Common Language Runtime garbage collection decides to free up the memory and destroys the class instance. However, between the time your code leaves the block until the Common Language Runtime destroys the class instance, you cannot access the members.

The block of code may be the block that encloses the code in a function, in which case the variable has *function scope*. When a function ends, variables that you have declared within the function no longer are in scope and you cannot access them.

In the *Account.cs* example, the variable *savings* has function scope. Only statements within the *Main()* function may access the variable. If you wrote another function within the *clsMain* class, that new function would not be able to access the variable within the *Main()* function.

Where you declare the variable makes a difference. You could have declared *savings* with block scope by enclosing it in the block of code that follows the *while* statement, and the results would have been radically different. Modify the *Account.cs* code and change the point where you declare *savings* as shown in the following:

```
static void Main ()
{
    while (true)
    {
        clsAccount savings = new clsAccount();
        // Other statements
    }
}
```

The *Account1* directory on the www.onwordpress.com Web site for this book contains these modifications to the *Account.cs* source-code file.

Compile and run *Account.cs* again. Each time the loop starts again, the program creates a new *savings* variable, and the value *Balance* is always its starting value, $0.00. The old variable goes out of scope and you no longer can access the new value that you set in later code.

The member fields and methods within the variable have *class scope*. You may access them only while the instance of the class exists and is within scope. Class scope is the largest scope you may have in C# because C# does not let you declare variables or functions outside of a class definition.

The scope limits do not apply to members that you declare *static*. Remember from Lesson 7, "Getting Started with Classes," that *static* members belong to the class, but non-*static* members belong to the instance.

WHAT YOU MUST KNOW

In this lesson, you learned how the access keywords can control and limit the range of statements that may access members of a class. You learned how to protect members from accidental access and how to prevent statements in other classes from giving member fields an improper value. The idea of placing access controls on members is part of the C# concept of *encapsulation*. In Lesson 12, you will learn how to derive new classes from a *base* class using the principle of *inheritance*. Although derived classes inherit the members of a base class, you still may want to limit access to base class members using the access keywords. Before you continue with Lesson 12, however, make sure you have learned the following key concepts:

- C# defines five levels of access that you may assign to fields, properties and methods that are members of a class.

- The *public* keyword is the least restrictive access, and functions in any other class may access *public* members of a class.

- Generally, you should try to keep your member fields *private* or *protected* and provide *public* member methods for other classes that need access to the fields.

- The range of statements that may access the members of a class is the *scope* of the variable that you declare to contain an instance of a class.

- Class members have *class scope*, and you may access the members only when an instance of a class is in scope.

DERIVING A NEW CLASS FROM AN EXISTING CLASS

In Lesson 8, "Object-Oriented Programming and C#," you examined *inheritance* and *polymorphism*—two object-oriented programming concepts that go hand-in-hand. Though inheritance, a new class may include the methods, properties and fields of another class, and, with some restrictions, use those members as though you had declared them as members of the new class. Through polymorphism, you may override members of the other class and modify the way the class operates. In this lesson, you will use the classes you wrote in Lesson 11, "Understanding Class Scope and Access Control," to derive new classes that will be very similar but have some fundamental differences. By the time you finish this lesson, you should understand the following key concepts:

- A *base class* contains the methods, properties, and fields that may be common to a group of related classes.
- A class derived from a base class may use the code and variables that you already have placed in the base class.
- When you create an instance of a derived class, the constructor for the derived class executes first, then the constructor for the base class executes.
- A derived class may override the code for a method that you declare as *virtual* in the base class.
- You may prevent an accidental instance of a base class by declaring the class or a member method as *abstract*.

As you program more applications using C#, you will have occasion to write many classes. Often you will find that in a single project, some classes have properties and functions in common. When you see this trend in a project, you may find it easier to define the common elements of each class in a single class definition. You can use the class as a *base class* to define the common behavior. Then, you can derive the more specific classes, adding the code and properties that the new class needs.

The *Account.cs* program from Lesson 11 uses the *clsAccount* class to describe a generic account. Suppose you want to describe more specific accounts, say, a checking account and a savings account. Both will need methods to add to the account balance and to subtract from the account balance. You might, however, need a method in the checking account to adjust for bank fees, and a method in the savings account to add interest earned.

First, you would want to build the generic class to describe the account, using methods to deposit and withdraw from the balance. The following code shows how you might design the *clsAccount* class:

```csharp
class clsAccount
{
    private decimal pBalance;
    private decimal Balance
    {
        get
        {
            return (pBalance);
        }
        set
        {
            pBalance = value;
        }
    }
    protected bool Deposit(decimal deposit)
    {
        if (deposit < 0)
```

```
            return (false);
        Balance += deposit;
        return (true);
    }
    protected int Withdraw(decimal withdraw)
    {
        if (withdraw < 0)
            return (1);
        if ((Balance - withdraw) < 0)
            return (2);
        Balance -= withdraw;
        return (0);
    }
    protected void ShowBalance()
    {
        Console.WriteLine ("\nYour account balance " +
                           "is {0:c}", Balance);
    }
    protected decimal GetBalance()
    {
        return (Balance);
    }
}
```

The first thing you should notice about *clsAccount* is that all the members are either *private* or *protected*. You could declare an instance of the class, but you would not be able to access any of the members. Only member functions of the class can access the *private* class members, and only members of the class and *derived* classes may access the *protected* members.

You can use the *clsAccount* class, then, only as a base class, from which you derive other classes. To access the *clsAccount* class members, you would have to derive a new class from *clsAccount* and provide *public* members within the derived class to manipulate the base class members.

To show this, derive two new classes, *clsSavings* to represent a savings account and *clsChecking* to represent a checking account. The following program, *Accounts.cs*, shows how to implement the new classes:

```
namespace nsAccounts
{
    using System;
    class clsMain
    {
        static void Main ()
        {
            clsSavings savings = new clsSavings();
            clsChecking checking = new clsChecking();
            while (true)
            {
                Console.WriteLine ("\nSelect one of " +
                                    "the following:");
                Console.WriteLine ("\tFor checking " +
                                    "enter 1:");
                Console.WriteLine ("\tFor Savings " +
                                    "enter  2:");
                Console.Write ("Enter Your selection  " +
                                " (0 to exit): ");
                        string strSelection =
                                        Console.ReadLine ();
                if (strSelection.Length == 0)
                    continue;
                if ((strSelection[0] < '0')
                    || (strSelection[0] > '2'))
                {
                        Console.WriteLine("\nYou entered " +
                                        "an invalid value");
                            continue;
                }
                int iSel = Convert.ToInt32(strSelection);
                if (iSel == 0)
                {
                    Console.WriteLine ("\nGoodbye!");
                    return;
                }
                if (iSel == 1)
                {
                    checking.Transaction ();
                }
                else if (iSel == 2)
```

```
            {
                savings.Transaction ();
            }
        }
    }
}

class clsAccount
{
    private decimal pBalance;
    private decimal Balance
    {
        get
        {
            return (pBalance);
        }
        set
        {
            pBalance = value;
        }
    }
    protected bool Deposit(decimal deposit)
    {
        if (deposit < 0)
            return (false);
        Balance += deposit;
        return (true);
    }
    protected int Withdraw(decimal withdraw)
    {
        if (withdraw < 0)
            return (1);
        if ((Balance - withdraw) < 0)
            return (2);
        Balance -= withdraw;
        return (0);
    }
    protected void ShowBalance()
    {
        Console.WriteLine ("\nYour account balance " +
                    "is {0:c}", Balance);
    }
```

```csharp
    protected decimal GetBalance()
    {
        return (Balance);
    }
}

class clsSavings : clsAccount
{
    protected string strAmount;
    protected decimal Amount;

    public decimal Transaction ()
    {
        Console.Write ("\nThe beginning balance in " +
                       "your savings account is " +
                       "{0:c}\n",
                       GetBalance());
        while (true)
        {
            Console.WriteLine ("\nSelect one of " +
                                "the following:");
            Console.WriteLine ("\tFor deposits, " +
                                "enter   1:");
            Console.WriteLine ("\tFor withdrawals " +
                                "enter 2:");
            Console.WriteLine ("\tTo add interest " +
                                "enter 3:");
            Console.Write ("Enter Your selection " +
                            "(0 to quit): ");
            string strSelection = Console.ReadLine ();
            if (strSelection.Length == 0)
                continue;
            if ((strSelection[0] < '0')
                || (strSelection[0] > '2'))
            {
                    Console.WriteLine("\nYou entered " +
                                "an invalid selection");
                    continue;
            }
            int iSel = Convert.ToInt32(strSelection);
            if (iSel == 0)
                return (GetBalance());
```

```
            if (iSel == 1)
            {
                Console.Write("Enter deposit amount: ");
                strAmount = Console.ReadLine ();
                Amount = strAmount.ToDecimal();
                AddDeposit(Amount);
            }
            else if (iSel == 2)
            {
                Console.Write ("Enter withdrawal " +
                                "amount: ");
                strAmount = Console.ReadLine ();
                Amount = strAmount.ToDecimal();
                SubWithdrawal (Amount);
            }
            else if (iSel == 3)
            {
                Console.Write ("Enter interest " +
                                "amount: ");
                strAmount = Console.ReadLine ();
                Amount = strAmount.ToDecimal();
                AddDeposit(Amount);
            }
        }
    }

    public decimal AddDeposit(decimal deposit)
    {
        bool result = Deposit (deposit);
        if (!result)
            Console.WriteLine("Invalid deposit " +
                                "amount {0:c}", deposit);
        ShowBalance ();
        return (GetBalance());
    }

    public decimal SubWithdrawal(decimal withdraw)
    {
        int result = Withdraw (withdraw);
        if (result == 1)
            Console.WriteLine ("Invalid withdrawal " +
                                "amount {0:c}", withdraw);
```

```csharp
            if (result == 2)
                Console.WriteLine ("You do not have " +
                        "enough in your savings account " +
                        "to withdraw {0:c}", withdraw);
        return (GetBalance());
    }
}

class clsChecking : clsAccount
{
    protected string strAmount;
    protected decimal Amount;

    public decimal Transaction ()
    {
        Console.Write ("\nThe beginning balance in " +
                        "your checking account is " +
                        "{0:C}\n",
                        GetBalance());
        while (true)
        {
            Console.WriteLine ("\nSelect one of " +
                                "the following:");
            Console.WriteLine ("\tFor deposits, " +
                                "enter  1:");
            Console.WriteLine ("\tFor checks, " +
                                "enter    2:");
            Console.WriteLine ("\tFor bank fees, " +
                                "enter 3:");
            Console.Write ("Enter Your selection "  +
                            " (0 to quit): ");
            string strSelection = Console.ReadLine ();
            if (strSelection.Length == 0)
                continue;
            if ((strSelection[0] < '0')
                || (strSelection[0] > '2'))
            {
                    Console.WriteLine("\nYou entered " +
                                "an invalid selection");
                        continue;
            }
            int iSel = Convert.ToInt32(strSelection);
```

```
        if (iSel == 0)
            return (GetBalance());
        if (iSel == 1)
        {
            Console.Write("Enter deposit amount: ");
            strAmount = Console.ReadLine ();
            Amount = strAmount.ToDecimal();
            AddDeposit(Amount);
        }
        else if (iSel == 2)
        {
            Console.Write ("Enter check amount: ");
            strAmount = Console.ReadLine ();
            Amount = strAmount.ToDecimal();
            SubWithdrawal (Amount);
        }
        else if (iSel == 3)
        {
            Console.Write ("Enter bank fees: ");
            strAmount = Console.ReadLine ();
            Amount = strAmount.ToDecimal();
            SubWithdrawal(Amount);
        }
    }
}

public decimal AddDeposit(decimal deposit)
{
    bool result = Deposit (deposit);
    if (!result)
        Console.WriteLine ("Invalid deposit " +
                            "amount {0:c}",deposit);
    ShowBalance ();
    return (GetBalance());
}

public decimal SubWithdrawal(decimal withdraw)
{
    int result = Withdraw (withdraw);
    if (result == 1)
        Console.WriteLine ("Invalid check " +
                            "amount {0:c}", withdraw);
```

```
         if (result == 2)
             Console.WriteLine ("You do not have " +
                 "enough in your checking account " +
                 "to cover a check for {0:c}",
                 withdraw);
         return (GetBalance());
     }
   }
}
```

Notice that in *Main()*, you now declare two class instances, *checking* for the checking account and *savings* for the savings account. In addition, rather than handle the transactions in *Main()*, the program passes control to a member function in the proper class. You now are using *Main()* as an entry point to the class, and the class handles all the details of the transaction.

The program derives the *clsSavings* or *clsChecking* classes from *clsAccount*. When you declare *checking* and *savings*, each instance includes its own copy of *clsAccount*, each containing its own member functions and fields. To demonstrate this, run the program and switch between *savings* and *checking*. Enter different transactions for each. As you switch between the accounts, you will see that each class maintains its own *Balance* property.

Using Access Keywords for Base Classes

Normally, this book has tried not to compare C# with C++ unless the differences are so significant as to lead to confusion for readers who also program in C++. This is one of those cases.

Unlike C++, C# does not let you specify an access keyword for the base class when you derive a new class from it. For example, in C++ you could inherit a class and specify an access of *protected*, as shown in the following snippet:

```
class DerivedClass : protected BaseClass
{
     // Class definition
}
```

In this declaration, the derived class would inherit all the *public* and *protected* members of the base class as *protected* members of the new class. Members that are *private* to the base class would remain *private*.

In C#, this declaration would cause the compiler to generate and print an error to your screen. The C# language specification does allow you to use the pro- tected and private keywords when you define a class object. You may declare the class as *public*, which permits access to the class from any code module. Another code module could use a *public* class to derive a new class. You also may declare a class as *internal*, in which you may use the class definition only within the current code module.

So far, you have written your programs using only one code module. In Lesson 13, "Using References and Assemblies," you will learn how to write modules that you may use in other projects.

If you do not give a class an access keyword, it will default to *public* and any program module in your program may create an instance of the class. To protect a class so that you may use it only in the program file in which you declare it, you can set the access level to *protected*.

To avoid confusion, this book has not used access keywords on classes yet. For programs that you write in a single source-code file, you do not need to worry about the access keyword. However, you should consider the content of your program. The default access has changed between releases of Visual C#, so if you write a set of classes that you might want to use elsewhere as a module or ref- erence, you should use the *public* keyword to make it clear that you want exter- nal programs to access the classes.

Hiding Class Members

You write base classes to be *generic*. You try to find the elements shared by the classes you intend to derive from the base class. There will be times that a generic method, property, or field in a base class does not exactly fit your needs in a derived class.

When this happens, you may *hide* a member in a base class by redefining it in the derived class using the *new* keyword. Using *new* effectively overrides the declaration in the base class, and functions in the derived class will reference the new member. Of course, functions in the base class do not know about the redefined member and will continue to reference the base class member.

To qualify for the *new* keyword, the base class member must be *public* or *protected*. Functions in derived classes do not have access to base class *private* members anyway, and so you do not have to use the *new* keyword.

To override a member of a base class, redefine the member in the base class by writing the access keyword (it does not have to be the same as the access level of the base class member), the *new* keyword, the data type the function returns, followed by the parameter list and statements.

For example, if you have an account balance defined in a base class of type *decimal* (for currency calculations) and you must derive a class to represent a stock account, you might want to use a *double* type to calculate the balance to more than two decimal places. The following snippet shows how you would do this:

```
public class clsAccount
{
    protected decimal dBalance;
}

public class clsHomeDepotStock : clsAccount
{
    protected new double dBalance;
}
```

You can apply the *new* keyword to a property as well. Remember that you do not have to use *new* to redefine a *private* field, so if *dBalance* is a private field that you access through a *public* property, you must redefine the property only:

```
public class clsAccount
{
    private decimal dBalance;
    public decimal Balance
    {
        get
        {
            return (dBalance);
        }
        set
        {
            dBalance = value;
        }
    }
}
public class clsHomeDepotStock : clsAccount
{
    private double dBalance;
    public new double Balance
    {
        get
        {
            return (dBalance);
        }
        set
        {
            dBalance = value;
        }
    }
}
```

From within the derived class, you still may access the base class object by qualifying the name with *base* followed by a period, then the name of the object. The following program, *Stocks.cs*, uses *Balance* in both the base and derived class and shows that each property may hold a separate value. In the base class,

Balance holds the total value of the stocks expressed in currency notation. In the derived class, *Balance* holds the total number of shares to the nearest one-thousandth:

```
namespace nsAccounts
{
    using System;
    public class clsMain
    {
        static void Main ()
        {
            clsHomeDepotStock stock =
                    new clsHomeDepotStock();
            stock.Balance = 67.9020;
            stock.LastQuote = 252.280;
            stock.ShowBalance();
            stock.ShowBaseBalance();
        }
    }

    public class clsAccount
    {
        private decimal dBalance = 0;
        public decimal Balance
        {
            get
            {
                return (dBalance);
            }
            set
            {
                dBalance = value;
            }
        }
        public void ShowBaseBalance ()
        {
            Console.WriteLine ("Total stock value = {0:c}",
                            Balance);
        }
    }
    public class clsHomeDepotStock : clsAccount
```

```
{
    private double dBalance;
    public new double Balance
    {
        get
        {
            return (dBalance);
        }
        set
        {
            dBalance = value;
            base.Balance = (decimal) (LastQuote *
                                     Balance);
        }
    }

    private double dLastQuote;
    public double LastQuote
    {
        get
        {
            return (dLastQuote);
        }
        set
        {
            dLastQuote = value;
            base.Balance = (decimal) (value * Balance);
        }
    }
    public void ShowBalance ()
    {
        Console.WriteLine ("Total shares held = {0}",
                           Balance);
    }
}
}
```

You may override methods in the same way. Normally, however, when you design a base class, you will use the *virtual* keyword for a method that you intend to override. You will learn about the *virtual* keyword later in this lesson.

Understanding the Order of Constructors and Destructors

If you define a constructor function for your class, your program will call the constructor automatically when you create an object from the class definition. Similarly, if you define a destructor function, your program will call the destructor automatically when the Common Language Runtime destroys the object.

When you derive a class from a base class, and you define constructor functions in both the base and derived classes, your program will call the base class constructor first and then call the derived class constructor. This gives the base class an opportunity to initialize any fields or properties that might be needed by the derived class. If there are more than one ancestor (base) classes, the program will call the constructors from the furthest ancestor first and continue to the nearest descendent.

A C# program calls destructors in the reverse order it calls constructors. First, the derived class destructor executes, then the base class. If there is more than one generation in the hierarchy, the most distant ancestor executes last.

In the following program, *Order.cs*, the class tree spans three "generations." Each constructor and each destructor prints out a message, so you can see the order in which they execute:

```
namespace nsOrder
{
    using System;
    public class clsMain
    {
        static void Main ()
        {
            clsSecondDerived MyClass =
                                    new clsSecondDerived();
            Console.WriteLine ("\nCode would execute " +
                            "here\n");
        }
    }

    public class clsBase
```

```
{
    public clsBase ()
    {
        Console.WriteLine ("Base Constructor called");
    }
    ~clsBase ()
    {
        Console.WriteLine ("Base Destructor called");
    }
}
public class clsFirstDerived : clsBase
{
    public clsFirstDerived ()
    {
        Console.WriteLine ("Constructor in first " +
                           "derived class called");
    }
    ~clsFirstDerived ()
    {
        Console.WriteLine ("Destructor in first " +
                           "derived class called");
    }
}
public class clsSecondDerived : clsFirstDerived
{
    public clsSecondDerived ()
    {
        Console.WriteLine ("Constructor in second " +
                           "derived class called");
    }
    ~clsSecondDerived ()
    {
        Console.WriteLine ("Destructor in second " +
                           "derived class called");
    }
}
}
```

When you compile and run *Order.cs*, you should see the following output:

```
Base Constructor called
Constructor in first derived class called
Constructor in second derived class called

Code would execute here

Destructor in second derived class called
Destructor in first derived class called
Base Destructor called
```

Using Virtual Functions

You learned in Lesson 8, "Object-Oriented Programming and C#," that the virtual function is the primary mechanism through which C# supports *polymorphism*. When you declare a function *virtual* in a base class and override the function in a derived class, other functions in the base class that call the *virtual* function search the derived classes for an *override* function.

At first glance, you might think there is little difference between *virtual* functions and the *new* method of redefining a function you learned in the section on hiding class members in this lesson. Programmatically, however, the difference is greater than you might think.

To demonstrate the difference, compile the following program, *Poly.cs*:

```
namespace nsPoly
{
    using System;
    public class clsMain
    {
        static void Main ()
        {
            clsDerived MyClass = new clsDerived();
            MyClass.ShowDiff();
        }
    }
```

```csharp
public class clsBase
{
    public void ShowDiff()
    {
        Virtual ();
        NewFunc();
    }

    public virtual void Virtual()
    {
        Console.WriteLine ("Virtual in base class");
    }
    protected void NewFunc()
    {
        Console.WriteLine ("NewFunc in base class");
    }
}

public class clsDerived : clsBase
{
    public override void Virtual()
    {
        Console.WriteLine ("Virtual in derived class");
    }
    public new int NewFunc()
    {
        Console.WriteLine ("NewFunc in derived class");
        return (0);
    }
}
}
```

First, when you use the *virtual* declaration, your *override* function may not change the access level or the return type of the function. Using the *new* keyword, you may completely redefine the function and change the return type and access level. Of course, if you change the number or type of parameters, the function has a different signature and you cannot use the *new* keyword. When you run *Poly.cs*, you will see the following output.

> **Virtual in derived class**
> **NewFunc in base class**

When the compiler encounters a *virtual* function, it builds a table of functions for the *virtual* function. Your program does not decide which function to call until you actually load the program into memory. At that time, the operating system consults the virtual table and selects the function in the most recently defined class and fixes that function as the one to call. This allows you to use *virtual* functions in a *library* that you compile in advance. You can see in the preceding output that, although the calling function is in the base class, the *Virtual()* function that it actually called is in the derived class. This is the principle of *polymorphism*: By overriding the *virtual* function, the derived class can change the behavior of the base class.

With a normal function, the compiler fixes the address of the function in the program. When you use the *new* method of redefining a function, no polymorphism takes place. Notice in the above output that the base class function called the *NewFunc()* function in the base class. The base class has no knowledge of the *NewFunc()* function in the derived class.

WHAT YOU MUST KNOW

In this lesson, you learned how to write base classes that you design with the intention of deriving new classes using the base class definition. You learned how to restrict the use of a base class to the current code module by using the *internal* keyword. You also learned how to hide members in a base class by using the *new* keyword and how that differs from the *virtual* keyword that you learned about in Lesson 8. Finally, you learned how your program calls constructor and destruction functions in a hierarchy of base and derived classes. In Lesson 13, you will start using the concepts you have learned to build code modules—assemblies—and to incorporate code that you already have written and compiled into new projects. Before you continue with Lesson 13, however, make sure you have learned the following key concepts:

- ❌ Base classes contain the methods, properties and fields that are common to the classes that you will derive from the class.

- ❌ By declaring a base class *internal*, you may prevent code in other modules from deriving new classes from the base class.

- ❌ If the particular members of a base class do not match the needs of a derived class exactly, you may *hide* the members by using the *new* keyword on a member of the same name in the derived class.

- ❌ When you create an instance of a derived class, your program calls the constructors beginning with the most distant base class first, then works its way down to the most recently derived class.

- ❌ If you define destructors in your base and derived classes, your program calls the destructors in the reverse order, beginning with the destructor in the most recently derived class and working backwards to the most distant base class.

USING NAMESPACES AND ASSEMBLIES

your programs so far have been short applications that you write in a single source-code file and then compile to produce an executable file. When you start developing large applications, things are not that simple. Granted, you can write a large program in one source-code file, and the Internet contains many examples of program files containing 30,000 or 40,000 lines of code in a single file. Managing the code in such a large file can become difficult, and programmers often break their code down into many shorter files. C# is very flexible in the ways you can handle multiple source-code files. In this lesson, you will learn how to create a program from more than one source file, how to create *modules* that you may include in other programs and how to create *assemblies*, compiled files that your program may reference without including them in the executable file. By the time you finish this lesson, you should understand the following key concepts:

- You may compile more than one source-code file in the same command line. The C# compiler will put the files together and produce a single executable file with code from all the source-code files.

- To guard against duplicating class names in other source files, you use namespaces in the source-code files.

- Rather than recompile *all* of a project's files when only some of the files have changed, you may compile the individual files into *modules,* components that contain compiled code.

- When you use a module in your project, the C# compiler adds the compiled code in the module to your executable file.
- An *assembly* contains compiled code that your program may reference when you write the source code.
- The compiler does not add the compiled code in an assembly to your executable. Instead, the Common Language Runtime loads the assembly into memory when you run your program.
- Assemblies help you to reduce the total size of an executable program. In addition, you may use an assembly to share code with other programs, even programs on other computers.

Working with Namespaces

Your C# programming is about to take a turn. Up to now, you have been writing short programs in a single source-code file. You then compiled that single source file into an executable file that performed a single task, to show you how a particular keyword or element of C# works.

Programmers rarely have that luxury in the real world, however. Even a small application is going to contain hundreds—and often thousands—of lines of code. Large applications may contain hundreds of thousands of lines of code. Managing all this code in a single source file would be a formidable task, so programmers usually break their project down into pieces. Some of these pieces may become parts of a single executable file, while others might become independent code modules that load into memory when the program needs them.

The C# environment provides several methods you may use to join separate program files together. You may compile them into a single executable at the same time by entering the names on the command line. You may compile one or more source-code files to produce a module, which you may later join with a program when you compile another source file. Thirdly, you may create an *assembly*, a file containing compiled source code that your program will load into memory when you run your program. In this lesson, you will experiment with all three methods.

The Visual Studio and C# have a number of assemblies that you may use in your code. Many of these provide basic library operations such as file reading and writing and, as you have seen, access to the console device. Among these assemblies, Microsoft has defined more than 1,000 classes to help you in your programming.

With so many classes available to you, there is a high possibility that you might duplicate class names in an assembly. To avoid this, Microsoft makes extensive use of namespaces. As you have learned, a namespace is a container for the classes and other objects that you use in a C# program. The names of classes in one namespace may duplicate those in other namespaces.

There is one global namespace, which is unnamed. If you define a class outside a namespace, the C# compiler places the class in this global namespace.

Take some time now to review namespaces in Lesson 5, "Understanding a Few Key C# Basics," if necessary.

Creating Your Own Namespaces

Except for the very early example, the sample programs in this book have declared their own namespaces, although most of them just as easily could have used the global namespace. To avoid duplicating class names within your own code as you develop more and more modules and assemblies, you must get used to using namespaces.

There is no established standard for naming namespaces, except that the name should describe the contents. To avoid duplicating namespaces established by the Microsoft library, I prefix namespace names with "ns." This also helps to identify them readily as namespaces I have declared.

After you have defined a class in a namespace, you may reference the class in another namespace, but you must inform the compiler that you want to reference the class in the other namespace. You do this with the *using* keyword, followed by the namespace name in which the class is defined. The most common example you have seen throughout the examples is the *Console* class, which is part of the *System* namespace:

```
namespace nsMine
{
    using System;
    class clsMine
    {
        //  Some functions and fields here
    }
}
```

After the *using* statement, the compiler knows to look in the *System* namespace for any class names that it cannot resolve in the current namespace.

You also may use the same namespace in more than one source-code file. When you use a namespace in more than one file, the compiler puts the files together and you may reference a class in either file without using the namespace. You will learn more about working with multiple source-code files in the next section.

Using Multiple Source-Code Files

When a source-code file becomes too long, it might be more convenient to break it down into two or more files. It is not unusual for a single program to contain several hundred lines of code. Simply scrolling through the source code to find a particular item might become time consuming. When you use more than one file, editing and managing the code becomes easier.

If you use more than one source file for your program, you specify each file on the command line when you compile your program. The C# compiler will put the files together and generate a single executable file. The name of the executable file will be the same as the first file you specify, but with an extension of *.exe*.

The following program breaks the *Stocks.cs* program from the previous lesson into two separate source files. The first is *Multi.cs*, which contains only the *clsMain* class. The following is the listing for *Multi.cs*:

```
/*
    multi.cs — demonstrates the use of more than one source
               file in a C# program. Used with balance.cs.

    Compile this program with the following command line:
           csc multi.cs balance.cs
 */

namespace nsAccounts
{
    using System;
    public class clsMain
    {
        static void Main ()
        {
            clsHomeDepotStock stock =
                    new clsHomeDepotStock();
            stock.Balance = 67.9020;
            stock.LastQuote = 252.280;
            stock.ShowBalance();
            stock.ShowBaseBalance();
        }
    }
}
```

You should notice that the *nsAccounts* namespace is complete in this file; that is, there is a beginning brace where it begins and a closing brace where it ends. That does not mean that you cannot continue using the namespace, but it must be complete in each file. In addition, notice that the declaration that you are *using System* is within the namespace braces.

The second is *Balance.cs*, which contains the classes that define the account:

```
/*
    balance.cs — Second source file for multi.exe.
                 See multi.cs for compile line.
 */
namespace nsAccounts
```

```
{
    using System;
    public class clsAccount
    {
        private decimal dBalance = 0;
        public decimal Balance
        {
            get
            {
                return (dBalance);
            }
            set
            {
                dBalance = value;
            }
        }
        public void ShowBaseBalance ()
        {
            Console.WriteLine ("Total stock value = {0:c}",
                               Balance);
        }
    }
    public class clsHomeDepotStock : clsAccount
    {
        private double dBalance;
        public new double Balance
        {
            get
            {
                return (dBalance);
            }
            set
            {
                dBalance = value;
                base.Balance = (decimal) (LastQuote *
                                          Balance);
            }
        }

        private double dLastQuote;
        public double LastQuote
        {
```

```
        get
        {
            return (dLastQuote);
        }
        set
        {
            dLastQuote = value;
            base.Balance = (decimal) (value * Balance);
        }
    }
    public void ShowBalance ()
    {
        Console.WriteLine ("Total shares held = {0}",
                            Balance);
    }
  }
}
```

Once again, notice that the namespace *nsAccounts* is complete within the second file as well. In addition, notice the *using System* declaration within the namespace boundary. Even if you placed the *using* statement outside the namespace, you still would have to declare it in both files.

Compile the program using the following command line:

```
C:> csc multi.cs balance.cs   <Enter>
```

The compiler will create an executable, *Multi.exe*, based on the name of the first file. When you run the program, you should see the same output you got from the *Balance.cs* program in the last lesson.

If you have more than two source files you must include in your program, simply type all the file names on the command line. The order is not important except for the first file name.

Actually, you may compile the source code for two different independent programs to generate a single executable file. Remember that each program must

have a *Main()* function that defines the program entry point. The C# compiler does not care how many *Main()* functions you have in a program, but you must specify which one will be the entry function.

For example, suppose you create the following program. For now, call it *OldProg.cs*:

```csharp
namespace nsOldProgram
{
    using System;
    class clsMain
    {
        static public void Main ()
        {
            clsOldClass MyObject =
                new clsOldClass ();
            Console.WriteLine ("In nsOldProgram.Main");
            Console.WriteLine ("Var = {0}", MyObject.Var);
        }
    }

    class clsOldClass
    {
        private int pVar = 42;
        public int Var
        {
            get
            {
                return pVar;
            }
            set
            {
                pVar = value;
            }
        }
    }
}
```

A few months later you get another idea for a similar program, but your friends or co-workers already are using *OldProg.exe*. You can incorporate all of the code into your new program without changing any code in the old program. You write the new program, now called *NewProg.cs*:

```
namespace nsNewProgram
{
    using System;
    using nsOldProgram;
    class clsMain
    {
        static public void Main ()
        {
            clsNewClass MyObject =
                new clsNewClass ();
            Console.WriteLine ("In nsNewProgram.Main");
            Console.WriteLine ("Var = {0}", MyObject.Var);
            nsOldProgram.clsMain.Main ();
        }
    }

    class clsNewClass
    {
        private int pVar = 21;
        public int Var
        {
            get
            {
                return pVar;
            }
            set
            {
                pVar = value;
            }
        }
    }
}
```

Now you compile the two program files to create a single executable with the following command line:

```
C:> csc /m:nsNewProgram.clsMain NewProg.cs OldProg.cs
```

The result is a new program, *NewProg.exe*, that contains the code from both programs. Your program has two *Main()* functions, so you use the /m compiler option to tell the C# compiler which you want to use as the entry point, in this case the *Main()* function in *nsNewProgram* namespace.

You still can call the *Main()* function in case your old code needs to do any initialization. Just qualify it with the namespace and the class name in the old program file.

When you run *NewProg.exe*, you will see that *Main()* in *NewProg.cs* executes first. That *Main()* then calls the *Main()* function in *OldProg.cs*.

Creating and Using Modules

Breaking your program down into multiple source files allows you to manage the code in each file more easily than a single, large file. The disadvantage is that you must recompile each file every time you make a change to your program. If you have four files and you only change one, you must recompile the other three as well.

Unlike the Visual C++ compiler, which produces a *.obj* file for each source-code file, the C# compiler produces no such intermediate files. The finished compilation is the executable without the need for object (intermediate) files. As a result, recompiling the entire program wastes your time and the computer's processor time.

You may compile one or more files into an intermediate library by specifying that you want a *module* file as the result of the compilation. You do this by using the /target option on the command line. For example, to create a module from the *Balance.cs* file of the last topic, you would use the following command line:

```
C:> csc /target:module Balance.cs   <Enter>
```

or you can use the short form:

```
C:> csc /t:module Balance.cs   <Enter>
```

Instead of a file that you execute directly, the compiler will create a module file with an extension of *.netModule*. The result of the preceding command would be a library file named *Balance.netModule*. As with compiling multiple files into a single executable, you may include more than one source file in a module.

After you have the module file, you may include it in any other program without having to recompile it. To compile the *Multi.cs* program file and include the *Balance.netModule* library file you just created, you would use the following command:

```
C:> csc /addmodule:Balance.netModule multi.cs
```

The result will be a single executable file, *Multi.exe*, that will contain the compiled code from *Multi.cs* and the code in the *Balance.netModule* module.

Creating modules keeps you from having to compile source files unnecessarily. In addition, the intermediate files may be used in a *make file* to generate your executable. There is not room in this book to cover make files adequately, but you can consult any reference on C++ if you are interested in building large applications from the command line.

Understanding Assemblies

C# is a *component-oriented* language. The design of the language encourages you to write programs in parts, or components.

The MSDN help file describes a component as "any useful, general-purpose object" that you write. It might be in C#, Visual Basic or even C++. From a practical standpoint, nearly every program you create using C# is a component. A notable exception is the module, the *.netModule* file you created in the last topic specifying /t:module on the command line when you compiled a program. However, it is very easy to convert a module to an assembly. The reason for this is that it is possible to combine several modules into a single assembly.

The basic, identifiable block of code in the .NET environment (which includes C#) is the *assembly*. An assembly is *self-describing* because it not only contains the intermediate code used by the Common Language Runtime, but an assembly also contains all the other information the program needs to use the code.

The core of the self-describing property is the *manifest*, a data structure that contains the identity and version of the assembly, the names of all the files in the assembly, the names of other assemblies that the applications need and the security permissions that users will need to execute the assembly.

You have been using assemblies throughout your programming in C#. The executable file that you create when you compile a program is an assembly. You may incorporate the executable file directly into another program without having to recompile the original code into a library.

The .NET environment defines two types of assemblies: private and shared. By far, the most common type of assembly you will use is a private assembly. Shared assemblies consume some system resource, and Microsoft designed .NET so that creating a shared assembly requires a bit more effort than creating a private assembly.

In this lesson, you will concentrate on private assemblies. You will use most of the same processes to create a shared assembly later.

Working with Assemblies

In this lesson, you have built executables from multiple source-code files. Then, you learned how to create an intermediate *module* and add the module when you compile your program.

To use an assembly, you need only add a reference to it when you compile your program. A major restriction of using private assemblies is that the assembly must be in the same directory as your program executable. In addition, if you create an assembly from modules, each module in the assembly must be in the same directory.

To add an assembly when you compile a program from the command line, you use the /r: compiler flag, followed by the name of the assembly. From within Visual Studio, the process is almost too easy to be true, and you will look at that shortly.

For the simplest example of using an assembly, return to the earlier example where you joined a new program, *NewProg.cs*, with an older one, *OldProg.cs*, simply by recompiling the source files into a new program. Without changing the original program files, add the following source file, *Assembly.cs*, to the same directory:

```csharp
namespace nsAssembly
{
    using System;
    using nsOldProgram;
    using nsNewProgram;
    class clsMain
    {
        static public void Main ()
        {
            clsAssembly MyObject = new clsAssembly ();
            Console.WriteLine ("In nsAssembly.Main");
            Console.WriteLine ("Var = {0}", MyObject.Var);

            nsNewProgram.clsMain.Main ();
        }
    }

    class clsAssembly
    {
        private int pVar = 84;
        public int Var
        {
            get
            {
```

```
        return pVar;
    }
    set
    {
        pVar = value;
    }
  }
 }
}
```

Now, compile *Assembly.cs* and add the original executable, *NewProg.exe* as a reference. Use the following command line:

```
C:> csc /r:newprog.exe assembly.cs   <Enter>
```

C# Assemblies

C# is a component-oriented language. The idea is that you can write a program in re-usable modules that your program may load into memory as it needs them. When you compile a C# program, the compiler writes your code into a package called an assembly.

In the .NET framework, the assembly is the basis for applications. It is convenient to think of an assembly as a library file, but it goes beyond simple functions that you place in a runtime library file. An assembly is a self-describing program file. It contains type information for the classes that you add to your program file, the Intermediate Language code and information on the resources you program uses.

What makes the assembly work is the manifest, a block of data about the assembly that compiler creates when you compile your program. The .NET framework will not execute any assembly that does not contain a manifest. The manifest contains the assembly name, a version number, information on the language (such as English) the assembly supports, and a list of all of the modules that you used to build the assembly as well other assemblies it will need to run properly.

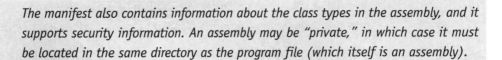

The manifest also contains information about the class types in the assembly, and it supports security information. An assembly may be "private," in which case it must be located in the same directory as the program file (which itself is an assembly).

An assembly also may be "shared." To make a shared assembly, you must add an encryption key to it, and place it in the global assembly cache. The encryption key prevents other programs from using the assembly unless they also have the encryption key. The Visual Studio contains all of the tools you will need to do this. Later in this lesson, you will learn how to use the tools to generate a key, add it to an assembly and then place the assembly in the global assembly cache.

Notice that the *Main()* function in *Assembly.cs* contains a call to the *Main()* function in *NewProg.cs*, which in turn calls the *Main()* function in *OldProg.cs*. When you run the *Assembly.exe* program, you should see the following output:

```
In nsAssembly.Main
Var = 84
In nsNewProgram.Main
Var = 21
In nsOldProgram.Main
Var = 42
```

The value of *Var* is different for each *Main()* function, showing that the old program did execute. It is important that you realize that *NewProg.exe* is *not* a part of your new *Assembly.exe* program. You have simply added a reference to the old program. When you run *Assembly.exe*, it loads *NewProg.exe* into memory and executes the functions in the program file.

One disadvantage of using an executable file as an assembly is that a user may execute it directly. It is handy during the development phase because you can test the program independently. When you *deploy* the project, however, you may not want a separate *.exe* file. You can change it simply by recompiling the old program file into a *library* file.

Return to the program directory and delete the *NewProg.exe* program file. Use the following line to compile *NewProg.cs* and *OldProg.cs* as a library file:

```
C:> csc /t:library NewProg.cs OldProg.cs  <Enter>
```

Notice here that you did not have to specify which *Main()* you want to be the entry point. A library needs an *.exe* file to run it, and you already have set the entry point in the *.exe* file. Notice that the compiler has created *NewProg.dll* rather than *NewProg.exe*.

You should notice also that you do not have to recompile *Assembly.cs* to incorporate the library file. Remember that an assembly is self-defining and contains all the information the program needs to access the code it contains. Of course, if you later recompile *Assembly.cs*, you will have to specify the *NewProg.dll* as the reference file rather than *NewProg.exe*.

You read earlier that a *module* that you create using the /t:module compiler flag cannot be used directly as an assembly. The primary use for a module is to link the code into your program file. The Visual C# toolbox contains a program, *AL.exe*, that will convert one or more module files into an assembly. When you do this, the module code file and the assembly file will be separate, and both must be in the same directory as the *.exe* file that uses them.

To show how this works, delete the *NewProg.netModule* file from the program directory. Use the following line to recompile the source files:

```
C:> csc /t:module NewProg.cs OldProg.cs  <Enter>
```

Notice that the compiler once again has created *NewProg.netModule*. If you try to use this file as an assembly, the compiler will tell you that it is not an assembly and you should use the /addmodule option.

To use the *AL.exe* program, you must specify an output file name for the assembly, and it must be different from the name of the file you are converting. Use the following command line to convert *NewProg.netModule* into an assembly.

```
C:> al /out:Prog.dll newprog.netModule  <Enter>
```

You should be aware that *Prog.dll* is an assembly that contains information about *NewProg.netModule*. For it to work, *both* files must be present in the program directory. Now recompile *Assembly.cs* to reference *Prog.dll* rather than *NewProg.netModule*:

```
C:> csc /r:prog.dll assembly.cs  <Enter>
```

Run the *Assembly.exe* program and you should see the same output as before.

You can use *AL.exe* to create an assembly that contains information about several modules. Assume you have compiled three different program files into modules—call them *Mod1.netModule*, *Mod2.netModule*, and *Mod3.netModule*. To create an assembly that references these modules, use the following command line:

```
C:> al /out:ProgAsm.dll mod1.netModule mod2.netModule
mod3.netModule  <Enter>
```

It is important that you remember that all four files—*ProgAsm.dll*, *Mod1.netModule*, *Mod2.netModule,* and *Mod3.netModule*—all must be in the program directory for the assembly to work.

Creating a Shared Assembly

Now that you have created private assemblies, you must experience the process of creating a shared assembly. To create a shared assembly, you must use cryptographic keys to guarantee that the names in an assembly are unique. The process actually is not complicated, and .NET contains all the tools you will need.

The advantage of using a shared assembly is that it does not have to be in the same directory as the program file. Instead, you can place it in the *assembly cache*, and the Common Language Runtime will look for it there.

First, make sure you have a module from which you will create the assembly. Use the following command line on the previous samples to generate the *NewProg.dll* module:

```
C:> csc /t:module NewProg.cs OldProg.cs   <Enter>
```

Next, you must generate a public and private key pair for the assembly. Only programs that "know" the assembly's private key may use a shared assembly. To generate the key pair, you use the shared name utility, *SN.exe*, using the *–k* flag to generate a key. You must also specify a file in which the utility will write the key pair:

```
C:> sn –k NewProg.snk   <Enter>
```

The key file may be any name you want to use. The extension for the key pair file does not have to be *.snk*, but that is one of the few conventions that has evolved from C#.

Next, you must generate the assembly using the key file. This will place the private key into the assembly file so that your program can access the assembly. You also should give it a version number. If you do not specify a version, it will show up in the assembly cache as version 0.0.0.0. Generate the shared assembly using the *AL.exe* utility with the following command line:

```
C:> al /keyfile:NewProg.snk /version:1.0.0.0 /out:ProgAsm.dll
NewProg.dll   <Enter>
```

You have generated a shared assembly. Now you must add it to the assembly cache using another utility, *GacUtil.exe*. You add an assembly to the cache using the *–i* flag and remove it using the *–u* flag:

```
C:> gacutil —i ProgAsm.dll   <Enter>
```

Now that the shared assembly has a key pair in it, you need to recompile your program, specifying the shared assembly name as a reference. For this step, the shared assembly file must be in the same directory that you use to compile your program:

```
C:> csc /r:ProgAsm.dll assembly.cs   <Enter>
```

You now can copy your program file *Assembly.exe* to another directory without having to copy the assembly file with it.

Now that you have learned how to place an assembly into the assembly cache, you probably want to know what the cache is. Basically, the cache is a *hidden* directory in your Windows directory, usually C:\Windows on Windows 95 and 98 systems and C:\WINNT on Windows NT and 2000 systems. You can view it using Windows Explorer.

Start Windows Explorer and navigate to the Windows directory. Look for a directory named *Assembly* and double-click the mouse on it. You should see the assembly cache complete with your *ProgAsm* assembly as shown in Figure 13.1. Notice that Explorer changes the header at the top of the list part of the window to display information about the assemblies.

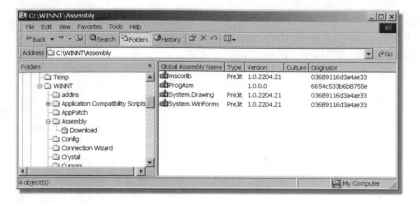

Figure 13.1. You can view the assembly cache using the Windows Explorer.

You also may add an assembly to the cache using drag and drop from a directory window to the cache window. If the assembly you are trying to drop into the cache is not a shared assembly, the drag-and-drop operation will fail.

To remove an assembly from the cache window in Windows Explorer, right-click your mouse on the assembly name and select Delete.

The purpose of all these steps is to keep you from reusing the code in an assembly that a commercial software product might install on your computer. Unless you know the private key—and a commercial software house is not likely to provide you with the key—you will not be able to write your own programs that use the shared assembly.

WHAT YOU MUST KNOW

In this lesson, you learned the importance of using namespaces when you write C# programs. You learned how to create your own namespaces and share them across multiple source-code files. You also learned how to compile a program using more than one source-code file, and how to create and use modules when you are compiling and building your program. Finally, you learned about private and shared assemblies and how to create and use assemblies. In the case of shared assemblies, you learned how to create a key file and add it to the assembly, and how to add the assembly to the system assembly cache. In Lesson 14, "Using Structures in C# to Group Related Data," you will continue your study of C# objects by learning about the *structure*. Before you continue with Lesson 14, however, make sure you have learned the following key concepts:

- To keep your code in manageable pieces, you may write your C# program in multiple source-code files.

- To guard against duplicating identifiers in different parts of your code, you may use namespaces. In programming with C#, the namespace is an important concept.

- When you use multiple files for one program, you may compile the individual source-code files into modules. You then may link the compiled modules into your program code.

- When you use a module, the compiler includes the module's code in your finished executable program. This increases the size of the executable.

- Instead of modules, you may create assemblies to hold your library code. The compiler does not add the assembly's code to your program. Instead, an assembly loads into memory when your program needs the code.

- When you use a private assembly, the assembly library file must be in the same directory as your executable file.

- Using a shared assembly, you may place your assembly in the system's assembly cache. Once in the cache, the assembly file does not have to be in the same directory as the executable file.

USING STRUCTURES IN C# TO GROUP RELATED DATA

O ver the course of the previous lessons, you have learned how to use the *class* data type in C# programs. You learned how to add fields and properties to the class, and how to add methods to manipulate the fields and properties. When you create an instance of the class, your program reserved memory on the heap to store the data members in the class instance. Thus, the variable containing the class instance is a *reference*-type variable. The class is the primary object type in C# programming, but it is not the only object type. In this lesson, you will learn about another C# data type that is very similar to the class, the *structure*. There are some important differences between a class and structure, but there are some marked similarities as well. By the time you finish this lesson, you should understand the following key concepts:

- The C# *structure* is a mechanism that lets you group related variables of different types together in a single object.

- The structure is very similar to the C# class. The structure, however, is a *value-type* object and the class is a *reference-type*.

- After you define a structure, you may create instances of the structure to store information in the member variables.

- Your program normally creates instances of a structure as value-type variables. However, when necessary you may use a structure variable as a reference-type.

- Structures may contain functions, including constructor functions. However, a structure may not contain a destructor function.

Defining Structures

Variables—fields and properties—use a memory location to store a value associated with the variable. In C#, the class is the *primary* container for variables as well as the functions—methods—that you write. It is not, however, the only container for variables and functions.

In C#, in addition to the class, you may define variables and functions in a *structure*. The structure is very much like a class, but, as you will see, the C# compiler and your program treat the structure very differently from the class.

You define a structure in the same way you define a class, except you use the keyword *struct* instead of *class*. First, you write the keyword *struct* followed by the name of the structure. You then begin the body of the structure with an opening brace, define your variables and functions, then end the definition with a closing brace:

```
struct MyStructure
{
    int     Var1;
    //  Other variable declarations
    void Func()
    {
        //  Function statements here
    }
    //  Other function definitions
}
```

As with the class, you do not need to end the structure definition with a semicolon, but the C# compiler does not seem to care either way. In the C and C++ languages, you must end the definition with a semicolon, so if you have experience in those languages you may prefer to use the semicolon.

Defining a structure does not create a structure object. It simply declares a new C# user data type. To create an instance of a structure, you must declare a structure variable, the same as you would for any other data type such as *int*, *double* or *char*. Notice in the following listing that you do not need to use the *new* operator as you would when you declare a class object:

```
namespace nsStructure
{
    using System;
    public struct SALE
    {
        public string  strItem;
        public string  strDescription;
        public int     iQuantity;
        public decimal dPrice;
    }

    class clsMain
    {
        static public void Main ()
        {
            SALE stSale;
            stSale.strItem = "Widget";
            stSale.strDescription = "That flippy thing";
            stSale.iQuantity = 24;
            stSale.dPrice = (decimal) 19.95;
            ShowItem (stSale);
        }
        static public void ShowItem (SALE Sale)
        {
            Console.WriteLine ("{0,-12}{1,-20}" +
                    "{2,5} @ ${3}    ${4:####.00}",
                    Sale.strItem, Sale.strDescription,
                    Sale.iQuantity, Sale.dPrice,
                    Sale.iQuantity * Sale.dPrice);
        }
    }
}
```

You should notice how similar the structure is to a class. To access a member of the object, you write the variable name, the *member* operator (a period) and the member name. You must set the access for members of the structure just as for a class. (If you have studied C++, you learned that the default access for structure members is *public*; in C#, the default access is *private*, the same as for a class).

Declaring Structure Members

A variable is a memory location in which your program stores a value. The C# structure lets you group values, even values of different types, using a single variable name.

To declare variables as members of a structure, you must write them as part of the structure definition. You write the declarations between the opening and closing braces that make up the body of the structure.

By default, the members of a structure are *private*, as with class members. Usually, however, you will want the members to be *public*, which is the default in C++. A structure encapsulates one or more variables, and there is little to gain by leaving the members *private*. To declare a variable as a structure member, first write the access keyword, then the data type, and finally the name you want to use for the member variable:

```
public struct SALE
{
    public string   strItem;
    public string   strDescription;
    public int      iQuantity;
    public decimal dPrice;
}
```

Unlike classes, you may not initialize a structure member when you declare it. The following would cause the C# compiler to generate an error:

```
public struct MyStruct
{
    public int       iAmount = 0;
}
```

As you learned, simply writing the *definition* of a structure only creates a new data type in C#. To create instance of the variables, you must declare a variable of the structure first. In the following, you use the structure to store the baud rate for your modem as a string (which you may display on your screen) and as a number:

```
public struct BAUDRATE
{
    public string   strRate;
    public long     lRate;
};

BAUDRATE B9600;
B9600.strRate = "9,600";
B9600.lRate = 9600;
```

After you have declared the *B9600* variable, you may set its values. The variables do not exist until you declare an instance of the structure.

Using Structures as Objects

The C# structure lets you declare multiple values, even values of different data types, using only a single variable name. When you create an instance of a structure, you create instances of all the variables that you have defined as members of the structure. This gives your program the ability to handle large or complex blocks of data more easily.

When you define a structure, you create a new C# data type. After you define the structure, you may declare new variables using the structure name as a data type. You may include structure variables in other structures, or as members of a class. To reference the member variables, you use the *qualified name* of the variable by writing the name of the structure variable, the member operator (a period) and the member name:

```
public struct MYSTRUCT
{
    public int iVal;
}

MYSTRUCT MyVar;
MyVar.iVal = 42;
```

In the following, you declare a structure named *POINT* that contains the x and y coordinates of a point on the screen. A *POINT* variable might be handy for storing the endpoints of a line, or the center of a circle. In this case, you define a variable of type *POINT* as a member of the structure *CIRCLE* to hold the coordinates of the center of a circle:

```
public struct POINT
{
    public int x;
    public int y;
}
public struct CIRCLE
{
    public CIRCLE (int CenterX, int CenterY, int Radius)
    {
        m_Center.x = CenterX;
        m_Center.y = CenterY;
        m_Radius = Radius;
    }
    POINT m_Center;
    int m_Radius;
}
```

An instance of a structure is a *value-type* variable as opposed the *reference-type* that you create when you declare an instance of a class. In C#, the distinction is important. When you use a value-type variable, the contents—the value you store in the variable—is the dominant characteristic. When you pass a value-type variable to a function, your program makes a copy of the value, and the function may not modify the original value. With a reference-type variable, the methods—the functions that define the behavior of the variable—is the dominant characteristic. When you pass a reference-type variable to a function, your program gives the function access to the original variable, and the function may modify the original value.

Understanding the Differences Between Classes and Structures

In C++, the class and the structure have more similarities than differences. In many instances, programmers could use a structure almost interchangeably with a class. In C#, while there are many similarities, it is clear that the designers of C# intended the structure to serve more as a container for related variables than as an alternative to the class.

Unlike a class, an instance of a structure is a value-type variable. Normally, your program creates value-type variables on the stack and you may use the *sizeof* and address operators on value-type variables in an *unsafe* block of code. However, if you try to use the *sizeof* operator on a structure, the compiler will issue an error that you cannot get the size of a managed type.

Defining a structure, you have learned, creates a new user-defined data type. After you define a structure, you may declare an instance of a structure simply by writing the new data type followed by the variable name:

```
SALE stSale;
```

You *may not* initialize the members of a structure at the time you declare the new variable. This is a major departure from C++, in which you could initialize the members by writing the values in a set of braces:

```
SALE stSale = {"Widget", "That flippy thing", 24, 19.95};
```

This declaration will cause the C# compiler to issue an error. To initialize a structure when you declare the variable, you would have to write a constructor function for the structure, and then declare it using the *new* operator. Doing so, however, does not change the structure variable from a value-type variable to a reference-type variable. A couple of short programs will demonstrate the similarities and differences between classes and structures.

First, the following program, *Struct1.cs*, uses an instance of the structure *SALE* as a value-type variable.

```csharp
namespace nsStructure
{
    using System;
    public struct SALE
    {
        public string  strItem;
        public string  strDescription;
        public int     iQuantity;
        public decimal dPrice;
    }

    class clsMain
    {
        static public void Main ()
        {
            SALE stSale;
            stSale.strItem = "Widget";
            stSale.strDescription = "That flippy thing";
            stSale.iQuantity = 24;
            stSale.dPrice = (decimal) 19.95;
            Console.WriteLine ("Before quantity change");
            ShowItem (stSale);
            AdjustQuantity (stSale);
            Console.WriteLine ("After quantity change");
            ShowItem (stSale);
        }
        static public void ShowItem (SALE Sale)
        {
            Console.WriteLine ("{0,-12}{1,-20}" +
                    "{2,5} @ ${3}    ${4:####.00}\n",
                    Sale.strItem, Sale.strDescription,
                    Sale.iQuantity, Sale.dPrice,
```

```
                          Sale.iQuantity * Sale.dPrice);
        }
        static public void AdjustQuantity (SALE Sale)
        {
            Sale.iQuantity = 36;
            Console.WriteLine ("In AdjustQuantity");
            ShowItem (Sale);
        }
    }
}
```

When you compile and run *Struct1.cs*, you will see the following output:

```
Before quantity change
Widget          That flippy thing        24 @ $19.95    $478.80

In AdjustQuantity
Widget          That flippy thing        36 @ $19.95    $718.20

After quantity change
Widget          That flippy thing        24 @ $19.95    $478.80
```

You should notice that while the *AdjustQuantity()* function changes the *iQuantity* member variable from 24 to 36, the original value in *Main()* remains unchanged. When you pass a value-type variable to a function, your program makes a *copy* of the original variable and uses the copy in the function call. Thus, the function cannot change the original value of the variable. Now you decide that, for coding efficiency, you must add a constructor to the definition of the *SALE* structure. When you do this, you must use the *new* operator to declare an instance of the structure, as shown in the following program, *Struct2.cs*:

```
namespace nsStructure
{
    using System;
    public struct SALE
    {
        public SALE(string Name, string Desc,
```

```
                          int Quantity, decimal price)
        {
            strItem = Name;
            strDescription = Desc;
            iQuantity = Quantity;
            dPrice = price;
        }
        public string  strItem;
        public string  strDescription;
        public int     iQuantity;
        public decimal dPrice;
    }

    class clsMain
    {
        static public void Main ()
        {
            SALE stSale = new SALE ("Widget",
                                    "That flippy thing",
                                    24, (decimal) 19.95);
            Console.WriteLine ("Before quantity change");
            ShowItem (stSale);
            AdjustQuantity (stSale);
            Console.WriteLine ("After quantity change");
            ShowItem (stSale);
        }
        static public void ShowItem (SALE Sale)
        {
            Console.WriteLine ("{0,-12}{1,-20}" +
                    "{2,5} @ ${3}    ${4:####.00}\n",
                    Sale.strItem, Sale.strDescription,
                    Sale.iQuantity, Sale.dPrice,
                    Sale.iQuantity * Sale.dPrice);
        }
        static public void AdjustQuantity (SALE Sale)
        {
            Sale.iQuantity = 36;
            Console.WriteLine ("In AdjustQuantity");
            ShowItem (Sale);
        }
    }
}
```

When you compile and run *Struct2.cs*, you should see exactly the same output as with *Struct1.cs*.

Finally, in *Struct2.cs*, change the definition of *SALE* from *struct* to *class*:

```
public class SALE
```

Classes and structures are similar enough that your program still should compile without any errors. However, when you run the program, you will see the following output:

```
Before quantity change
Widget          That flippy thing      24 @ $19.95    $478.80

In AdjustQuantity
Widget          That flippy thing      36 @ $19.95    $718.20

After quantity change
Widget          That flippy thing      36 @ $19.95    $718.20
```

Notice particularly the last line of the output. The *AdjustQuantity()* function has changed the original value of the *iQuantity* member. When you use a structure, you pass a *copy* of the variable in the function call and changes to the structure do not change the original structure. When you use a class, however, you pass a *reference* to the original object rather than a copy and changes to the class members also change the original values.

You should understand that if you provide a constructor function for a structure, the parameter list may not be empty. The following constructor will compile properly for a class definition, but would cause a compiler error in a structure definition:

```
public SALE()
{
    strItem = "Widget";
    strDescription = "That flippy thing";
```

```
    iQuantity = 24;
    dPrice = (decimal) 19.95;
}
```

You cannot define a constructor with an empty parameter list because C# provides a default constructor for structures. The default constructor takes no parameters, and so your constructor would duplicate the default constructor. Typically, the default constructor will initialize numeric members to 0 and string members to an empty string (an empty string is one that contains no characters).

You should be aware of a couple of other differences between the C# class and a structure:

❶ A structure may not contain a destructor function. This is a major difference between C# structures and C++ structures.

❷ You may not use a structure as a base to derive a new structure. Thus, a structure does not use the inheritance principle. This also is a major difference between C# and C++.

Treating a Structure Variable as a Reference-Type Object

Although variables you create from a structure definition are value-type variables, there are times you might want to use them as reference-type objects, and you want a function to be able to modify the original values. You may temporarily modify a value-type variable to a reference-type variable by using the *ref* keyword in front of the variable's name.

The following program, *Circle.cs*, adds properties and methods to the *CIRCLE* structure. After you define the center point and the radius, you use the *Area* property to calculate the area of the circle, and the *PointOnCircle()* method to calculate the screen coordinates of a point on the circle. The coordinates of the point on the circle use two values, one for the x coordinate and another for the y coordinate. A method, however, may return only a single value. To overcome this, in the *Main()* function, you use the *ref* keyword to change the *point* variable to a reference-type. The *PointOnCircle()* method then may change the original values:

```csharp
namespace nsCircle
{
    using System;

    public struct POINT
    {
        public int x;
        public int y;
    }
    public struct CIRCLE
    {
        public CIRCLE (int CenterX, int CenterY, int Radius)
        {
            m_Center.x = CenterX;
            m_Center.y = CenterY;
            m_Radius = Radius;
        }
        POINT m_Center;
        int m_Radius;
        public double Area
        {
            get
            {
                double radius = (double) m_Radius;
                return (radius * radius * 3.14159);
            }
        }
// Use a reference to return the two values in the
// point parameter.
        public void PointOnCircle (int Angle,
                                    ref POINT point)
        {
//          Convert the angle from degrees to radians
            const double Radian = 57.29578;
            double fAngle = ((double) Angle/10.0) / Radian;
            point.x = (int)((double) m_Radius
                        * Math.Cos (fAngle)) + m_Center.x;
            point.y = (int)((double) m_Radius
                        * Math.Sin (fAngle)) + m_Center.y;

        }
```

```
    }
    public class clsMain
    {
        static public void Main ()
        {
            CIRCLE circle = new CIRCLE (20, 42, 200);
            POINT point;
            point.x = point.y = 0;

// Temporarily make the point variable a reference-type
            circle.PointOnCircle (450, ref point);
            Console.WriteLine ("The area of the " +
                            "circle is {0}",
                            circle.Area);
            Console.WriteLine ("The point on the circle " +
                            "at angle 45.0 is ({0}, {1}),
                            point.x, point.y);

        }
    }
}
```

You should notice that you must write the *ref* keyword in the parameter list when you write the *PointOnCircle()* function, and again when you pass the *point* variable in *Main()* to the function.

Try removing the *ref* keyword from both the *PointOnCircle()* function and the statement in *Main()* that calls the function. You will see that without the *ref* keyword, the function is unable to modify the original values, and the *Console.WriteLine()* function prints *(0, 0)* as the location of the point.

WHAT YOU MUST KNOW

In this lesson, you learned about the C# structure, a class-like construction that your program may use to declare value-type variables that may contain multiple values, methods and properties. You learned the difference between the C# class and the structure, and how you may use a structure to encapsulate multiple values that your program may pass to a function using a single variable. A function cannot change the values of the original structure when you pass it as an argument. However, you also learned how to override this behavior to pass a structure variable as a reference-type variable, thus allowing a function to modify the original values. In Lesson 15, "Handling Exceptions," you will learn how to handle unexpected errors that occur in C# programs. Before you continue with Lesson 15, however, make sure you have learned the following key concepts:

- Using a *structure*, your C# program may declare multiple values—even values of different types—using a single variable name.

- The structure gives your program the ability to handle large and possibly complex groups of values using only one variable.

- A structure is similar to a class, but instances of a structure are *value-type* variables. Instances of a class are *reference-type* variables.

- When you include methods and properties in a structure definition, you must declare instances of the structure using the *new* operator.

- A structure may contain a constructor function but no destructor function. A constructor function may not have an empty parameter list.

- When necessary, you may temporarily change a structure variable to a reference-type variable by using the *ref* keyword.

HANDLING EXCEPTIONS

s your programs become more complex and begin to perform useful purposes, the chances increase that a program will encounter an unforeseen error. Often when this happens, C# and the Common Language Runtime will "throw an exception." In Lesson 6, "Using Variables to Store Information in C# Programs," you saw how a simple typing error such as entering the wrong parameter number for the *Console.Write()* function can cause your program to end abruptly. Exception-handling is important in C#, and you can "catch" and process these errors so that your program can recover and continue. In this lesson, you will learn how to throw and catch exceptions and how to process them. By the time you finish this lesson, you should understand the following key concepts:

- An *exception* is an interruption in the program flow caused by unexpected or unforeseen events.
- To handle an exception, the code that causes—or "throws"—the exception must be within a *try* block. When an exception occurs within a try block, your program transfers control to the appropriate *catch* block.
- In addition to natural events, you may force your program to generate an exception using the *throw* statement.
- To throw and catch exceptions, you must use one of the C# exception classes, or use a class that you derive from one of the C# exception classes.

- The *finally* block in an exception-handling block contains code that is guaranteed to execute, regardless of the results of the code in the *try* block.

Understanding Exception Handling

No matter how carefully you design and write your program, there is always the possibility that some unforeseen event will cause the program to encounter an error that makes it impossible to continue. The error may be the result of a typing mistake that did not cause the compiler to issue an error but nonetheless causes a program error. Or it may be something beyond your control. The computer on which your program is running may not have enough memory, you may run out of disk space, or a device attached to the computer might not be functioning properly. Many of these problems will cause your program to thrown an *exception*.

An exception is an interruption of the normal program flow resulting from an unforeseen or unexpected event. Exceptions are a part of the C# language and give the programmer an opportunity to detect and recover from these events. Your program can "catch" the exception, take whatever action is necessary and then either continue executing or exit gracefully. An *exception-handler* block allows you to provide an alternate block of code to execute if you catch an exception. If you do not catch and handle an exception, your program will end abruptly.

Your program may generate an exception under three conditions:

- You may choose to throw an exception to signal an error condition as a means of aborting a block of code. For example, if a file that you need for a block of code does not exist, you may throw an exception that will execute a block of code to create the file first.
- Your program may throw an exception as the result of a hardware problem, such as an attempt to divide by zero or to access an invalid memory address.
- Finally, your program may encounter a software problem that keeps it from completing a task, such as a request to allocate an invalid block of memory or an attempt to access memory through an invalid handle or pointer.

You test for an exception inside a *guarded* section of code. The guarded section begins with a *try* statement and contains the statements where you expect an exception. If your program throws an exception while executing statements in the guarded block, it transfers control to an alternate section of code, the *catch* block. Exception handling in C# is very similar to exception handling in C++ and Java. Unlike C++, however, you can only throw an exception using a class object derived from *System.Exception*; you cannot use a fundamental data type to throw an exception.

The .NET framework does contain some default exception handlers for common exceptions, such as divide by zero. The divide-by-zero exception is a hardware error generated by the *central processing unit* in your computer when a program tries to divide a number by zero. The following short program, *DivZero.cs*, performs such an operation in C#:

```
namespace nsDivZero
{
    using System;
    public class clsMain
    {
        public static void Main()
        {
            double x = 42.6;
            double y = 0.0;
            Console.WriteLine ("{0} / {1} = {2}",
                                x, y, x / y);

        }
    }
}
```

When you compile and run this program, you should see the following text on your screen:

```
42.6 / 0 = Infinity
```

Mathematically, of course, that is incorrect, but it does show that the runtime code handles the exception generated by the illegal division. It will keep your program from crashing if it performs this type of operation.

Throwing an Exception

To begin your study of exception handling, you must see what a typical user will see when your program throws an exception. However, when you installed Visual Studio.NET, by default you enabled *Just In Time Debugging*, or "JIT." When your program throws an exception, the runtime environment displays a dialog box that lets you enter the Visual Studio debugger to investigate the problem.

When you distribute your program, however, the typical user's computer is not likely to have JIT installed and enabled. You must disable JIT on your development computer to enjoy the full effect of an unhandled exception as seen by the user. To do disable the just-in-time compiler, perform the following steps:

❶ Within Visual Studio, select the Tools Menu and choose Options. Visual Studio will display the Options dialog box.

❷ Within the Options dialog box folder window, select the Debugging folder. If necessary, double-click your mouse on the Debugging folder and expand the folder to show its categories.

❸ Within the Debugging folder, select the Just-In-Time category. The Visual Studio will display the Just-In-Time Debugging dialog box.

❹ Within the JIT Debugging Settings dialog box, you can enable and disable JIT debugging for any program you develop using the Visual Studio. Because you are interested in the "Common Language Runtime" type, uncheck this item in the box that lists the program types.

❺ Click your mouse on the OK button to close the Options dialog box.

Your C# programs will no longer automatically start the Visual Studio debugger. To re-enable JIT, follow the above steps, but check the "Enable JIT Debugging" box.

The following program, *Except1.cs*, does nothing but throw an exception:

```
namespace nsExcept
{
    public class clsMyException : System.Exception
    {
    }
    public class clsMain
    {
        static public void Main ()
        {
            throw (new clsMyException());
            Console.WriteLine ("Exception was handled");
        }
    }
}
```

Compile and run *Except1.cs*. You should immediately see an exception dialog box similar to the one in Figure 15.1. The line that writes the message to the console never gets a chance to execute.

Figure 15.1 When your program throws an exception that it does not handle, Windows displays an exception dialog box.

The result is not very polite. Any unsaved work at the point where your program throws the exception will be lost. If the user were in the process of entering text or filling out a large form, you might expect a phone call from an upset user.

To handle the exception, you need to write a *try* block where you will write statements that may throw an exception. The try statement creates the guarded block. Next, you must write a *catch* block, where you will write alternate code that will execute only if your program throws an exception.

```
/*
    Except2.cs — demonstrates throwing and catching
                  an exception.
    Compile this program with the following command line:
          C:>csc except2.cs
 */
namespace nsExcept
{
    using System;
    using System.Windows.Forms;
    public class clsMyException : System.Exception
    {
        public clsMyException (string Message)
        {
            m_Title = Message;
        }
        public void ShowException (string Text)
        {
            MessageBox.Show (Text, m_Title);
        }
        protected string m_Title;
    }
    public class clsMain
    {
        static public void Main ()
        {
            try
            {
                throw (new clsMyException("Throwing " +
                                        "Exception"));
            }
            catch (clsMyException e)
            {
                e.ShowException ("Handled. Cleaning up " +
                                "and Exiting gracefully");
                //  Write clean up and exit code here
                return;
            }
            Console.WriteLine ("Exception was handled");
        }
    }
}
```

First, *Except2.cs* defines a class, *clsMyException*, which is derived from *System.Exception*. In the *throw* statement, you must declare an instance of a class derived from *System.Exception*. (Actually, you could use *System.Exception*, but it would display the message box shown in Figure 15.1). The *clsMyException* class creates a custom message box as shown in Figure 15.2.

Figure 15.2 The exception class provides a custom message box to display the exception.

In *Main()*, you enclose the guarded code in the *try* block. In this case, the only statement is the *throw* statement, but you could include virtually any code here. When your program throws the exception, control passes to the block of code following the *catch* statement.

The *catch* block uses the exception object to display a message box. After you click your mouse on the OK button on the message box, the *return* statement ends the program. Of course, the last line in the program—the *ConsoleWriteLine()* statement—still does not execute.

You do not have to return from the *catch* block. If you remove or comment out the *return* statement, the program would execute the code in the catch block and then continue normally.

What to Do When C# Throws Exceptions

Exception handling offers a number of advantages for your program. Before exception handling became a part of the C++ programming language, functions, including library functions, signaled an error by returning an error code. It then was up to the programmer to detect and act upon the error code.

This presented some problems. The programmer could simply ignore the error code, even if the value indicated a severe system problem that would prevent the program from continuing normally. In addition, some functions are *void* type and cannot return a value. Class constructors, for example, do not have a return type and thus cannot return an error code. If your class instance requires a key operation such as opening a file or allocating memory, the constructor cannot return a code signaling failure.

When a program throws an exception, however, the program cannot ignore the exception. If the program code does not handle the exception, the program ends. In addition, a *void* type function or a class constructor can throw an exception, thus signaling an error where it is otherwise impossible to return an error code.

C# and the Common Language Runtime use exceptions extensively, and in many situations where C++ does not. For example, if you create an array with 20 elements and then try to access the 25th element, the Common Language Runtime will throw an exception using the *IndexOutOfRangeException* class. C++ will not throw any exception in this case.

It is important that you understand at this point that the object in a *catch* statement must be of the same type used when the program threw the exception. If your program throws a *clsMyException* object, the catch statement must have an argument of type *clsMyException*. It is not an error if the types are different. The program just will not catch the exception unless it is of the same type. The following short program, *Except3.cs*, declares two exception classes, and then throws the one the program is not expecting:

```csharp
namespace nsExcept
{
    using System;
    class clsExcept1 : System.Exception
    {
    }
    class clsExcept2 : System.Exception
    {
    }
    class clsMain
    {
        public static void Main ()
        {
            try
            {
                throw (new clsExcept2());
            }
            catch (clsExcept1 e)
            {
                Console.WriteLine ("Exception caught");
                return;
```

```
            }
        }
    }
}
```

When you compile and run *Except3.cs*, you will get the same rude error message that you would get if you did not try to handle the exception. Change the types so that they match and your program will catch and handle the exception. Later in this lesson, you will learn how to handle multiple exception types.

With this in mind, when you are writing code to catch specific errors in your program, you must use the proper exception class that a method might throw. C# provides a number of exception classes, as summarized in Table 15.1.

Class	Reason
AccessException	The program failed to access a type member, such as a method of field.
ArgumentException	Signals an invalid argument to a function.
ArgumentNullException	Signals that a null argument was passed to a function that cannot handle it.
ArgumentOutOfRangeException	Signals that an argument to a function is outside an acceptable range.
ArithmeticException	Signals arithmetic overflow or underflow.
ArrayTypeMismatchException	The program attempted to store the wrong type of object in an array.
BadImageFormatException	The format of a dynamic link library or executable file is incorrect.
CoreException	Base class for runtime exception classes.
DivideByZeroException	Attempt to divide by zero.
FormatException	The format of an argument is incorrect.
IOException	A read or write to a device failed.
IndexOutOfRangeException	The index for an array is out-of-bounds.
InvalidCastException	Signals an attempt to cast an object to an invalid class.
InvalidOperationException	A call to a method violated the rules for calling the method.
NotFiniteNumberException	A number value is invalid.
NotSupportedException	Attempt to call a method that a class does not support.
NullReferenceException	Attempt to dereference a null object.
OutOfMemoryException	Allocation attempt for a new object failed.
StackOverflowException	The stack pointer has exceeded its maximum value.
SystemException	Base class for other exception classes. Signals a failed run-time check.

Table 15.1 Exception classes used by C# and the Common Language Runtime.

To determine which exception you must catch, you can look up the method in the MSDN help file. From the Visual Studio, select Index on the Help menu, then type in the name of the method in the Look For field. The method description will tell you what exception class the method will throw and what namespace you need to use.

Understanding Scope in Exception Blocks

In Lesson 25, "Understanding Function and Variable Scope," you will learn about *scope*, an important characteristic of fields and properties. The scope of an identifier is the range of statements that may access the identifier, whether it is a field, a property or a method. However, you must know the basics of scope and how it relates to the exception blocks.

Depending upon how and where you declare a variable, only a limited group of statements will be able to use the variable. If you declare a variable inside a function, for example, the variable is local to the function and statements outside the function block are not aware the variable even exists. The variable has *function scope*. The scope of a variable within a function begins when you declare the variable. Statements before the point of declaration may not use the variable.

Variables also have a *block scope*. A block is any group of statements enclosed by a set of braces. You may declare a block of code anywhere within a function, and any variables you declare within the block will not be accessible to statements outside the block. A block also may be a *compound statement* enclosing the code for a conditional or loop statement. The following snippet declares a variable *x* in function scope, and a variable *y* in block scope:

```
void Func()
{
    int x;          // x has function scope
// Statements here may use x but not y.
    {
        int y;      // y has block scope
```

```
// Statements here may use either x or y
    }
// Statements here may use x but not y.
}
```

Statements outside the block may not access variables declared within the block. In addition, variables declared within the block exist only when your program is executing statements within the block.

At its basic, the scope of a variable is from the point of declaration to the end of the block that encloses it. When you use exception-handling blocks you should be aware that both the *try* and *catch* keywords require compound statements, and any variables that you declare within the compound statements follow the rules that determine the scope of variables.

Any variables that you declare within a *try* or *catch* block will go out of scope when the block ends. If you create a variable within a *try* block and an exception occurs, the variable will go out of scope when your program transfers control to the *catch* block. Similarly, any variables you declare within a *catch* block are not accessible to statements in the *try* block:

```
try
{
    int iVar = 0;  // declared in the try block
    // Statements in the try block.
}
catch (FormatException e)
{
    // The iVar variable in the try block is out of
    // scope and not available here.
    double fVar;   // declared in the catch block. fVar is
                   // not accessible from the try block
}
```

If you must use the same variable in both the *try* and *catch* blocks, you should declare the variable *before* entering the *try* block.

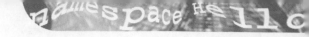

Exception-handling is specific. When you throw an exception, you must pass an exception object (a class instance that you create from an exception class). If the object expected by the *catch* statement is not the same type as the object you throw, the *catch* block will not execute.

When you write the *catch* block for the exception handler, you must include a parameter of the same data type as the type of value passed by the *throw* statement. If the data types are not the same, your *catch* block will not trap and process the exception:

```
try
{
    throw (new IOException());
}
catch (FormatException e)
{
    // This catch block declares a FormatException so it
    // will not catch the above exception, which passes
    // an IOException value.
}
```

There is no generic exception handler in C# as there is in C++. In C#, you may cast any class object to an object of a base class. If you have an object *obj* that is an instance of *clsClass*, which is derived from *clsBase*, you may cast *obj* to a *clsBase* object. You may use this technique to create a generalized exception handler. If you are not sure what exception class to use or you cannot find the information in MSDN, you can use this generic *Exception* class:

```
catch (Exception e)
```

Exception is the base class for all of the exception objects, and this throw statement will catch virtually all exceptions thrown by the Common Language

Runtime. When you do this, you can use the *Exception.ToString()* method to get the name of the exception the Common Language Runtime threw:

```
catch (Exception e)
{
    Console.WriteLine (e.ToString());
}
```

When you apply this to the formatting problem, you will get several lines of information about the exception class. The first line, however, tells you the actual exception class, in this case the *FormatException* class.

```
System.FormatException: The index (zero based) must be greater
than or equal to zero and less than the size of the argument list.
```

Although you used an *Exception* class object in the catch, C# tells you that it actually caught a *System.FormatException*.

Terminating a Program in an Exception Block

If a part of your program throws an exception and your code does not handle it in *try* and *catch* blocks, your program is going to terminate, and rather rudely from the user's standpoint. Any information in your program's memory will be lost, including perhaps some important information the user might have entered.

If the problem that caused the exception is so severe that your program cannot continue, you can use the *catch* block to terminate your program gracefully, saving any information or text the user might have entered.

This does not mean that every statement in your program should be protected by a *try* block. A statement such as $x = y / 2$ is not going to throw an exception. However, key operations that your program needs to function—such as accessing a hardware port or opening a file—should be guarded in a *try* block.

In the previous example, you have used the *return* statement to exit the *Main()* function, thus terminating your program. That is not always an option, however. An exception may be thrown in the code in a function that your program is executing, and you may choose to handle the exception in the function instead of *Main()*. If you simply execute a *return* statement, your program will return to *Main()* and continue running, as shown in the following short program, *ExcAbort.cs*:

```csharp
namespace nsAbort
{
    using System;
    class clsMain
    {
        public static void Main()
        {
            clsMain main = new clsMain();
            main.CauseProblem ();
            Console.WriteLine ("Application did not exit");
        }
        public void CauseProblem()
        {
            try
            {
                throw (new CFormatException());
            }
            catch (CFormatException e)
            {
                Console.WriteLine (e.Message);
                return;
            }
        }
    }
}
```

When you compile and run *ExcAbort.cs*, the text "Application did not exit" displays on your screen, indicating that the program did not exit in the *catch* block.

In C++, programmers may call a global function *exit()* to terminate a program at any point in the code. (A global function is a function that is not a member of any class.) C#, however, does not permit global functions.

The *Environment* class in the *System* namespace does contain a *static* function *Exit()* that will terminate an application, even in a function call. When you call the function, your program ends at the point where you execute the *Environment.Exit()* function call, and does not return to *Main()*.

Change the *CauseProblem()* function in *ExcAbort.cs* as follows:

```
public void CauseProblem()
{
    try
    {
        throw (new CFormatException());
    }
    catch (CFormatException e)
    {
        Console.WriteLine (e.Message);
        Environment.Exit (-1);;
    }
}
```

Recompile *ExcAbort.cs* and run the program again. The program will terminate when it executes the *Environment.Exit(-1)* statement. The *–1* is an *exit code* that the program returns to the operating system. By convention, programs return a 0 to the operating system when they exit normally, and a non-zero value when an error condition causes a program to terminate.

When you are writing programs using Windows forms, you may use the *Application.Exit()* method, which closes all windows before ending the program.

Using Multiple *catch* Blocks

You learned in this lesson that a *catch* block expects an object of the same type used in the *throw* statement. If the objects are not of the same type, the *catch* block will not execute, and the exception will cause your program to end.

Generally, it is not a good idea to use the base *Exception* class to catch all exceptions. Your program may throw an exception that you truly do not expect and which could cause your program some serious problems. Still, you will find instances where you need to handle more than one type of exception.

You can "stack" *catch* blocks simply by adding one right after another. Each block must expect a different exception object type, and you cannot have any statements between the *catch* blocks. Suppose your program executes a function that might throw a *FormatException* or a *DivideByZeroException*. The following program, *MultiExc.cs*, shows how you could handle both types:

```csharp
namespace nsMultiExcept
{
    using System;
    class clsMain
    {
        public static void Main()
        {
            clsMain main = new clsMain();
            try
            {
                main.CauseProblem ();
            }
            catch (FormatException e)
            {
                Console.WriteLine ("Format Exception");
                Console.WriteLine (e.Message);
            }
            catch (DivideByZeroException e)
            {
                Console.WriteLine ("Div zero Exception");
                Console.WriteLine (e.Message);
            }
        }
        public void CauseProblem()
```

```
      {
              throw (new DivideByZeroException());
      }
    }
}
```

When you use multiple exception blocks, your program will check each block for a matching type when it throws an exception. When it finds a matching type, that block will execute and the program will ignore all other blocks. If you use a generic exception handler, it should be the last in the sequence, as shown in the following snippet:

```
clsMain main = new clsMain();
try
{
    main.CauseProblem ();
}
catch (FormatException e)
{
    Console.WriteLine ("Format Exception");
}
catch (DivideByZeroException e)
{
    Console.WriteLine ("Div zero Exception");
}
catch (Exception e)
{
    Console.WriteLine (e.Message);
}
```

If you were to put the generic handler first, it would catch every exception, and the specific handlers would never execute.

If your program throws an exception that it does not handle in *try* and *catch* blocks, your program will execute the default exception handlers, which will cause your program to terminate after displaying an error message.

There may be times that you will want the default exception handlers to perform their job, however rude it might be, yet still provide some cleanup code before your program terminates. C# provides a mechanism to perform this action through the *finally* block.

You may use a *finally* block instead of a *catch* block when you write a guarded block of code, or you may use it in addition to one or more *catch* blocks. If you use it with *catch* blocks, the *finally* block must be the last block in the chain.

The *finally* block always will execute, regardless of the results of the code in the *try* block. If your program throws or does not throw an exception, the *finally* block will execute. If your program catches or does not catch an exception, the *finally* block will execute. In short, the *finally* block *will* execute (assuming you do not hit the Big Red Switch). Thus, when used with a *try* block, the *finally* block is guaranteed to execute.

You may use a *finally* block without any catch blocks, such as in the following:

```
try
{
    main.CauseProblem ();
}
finally
{
    Console.WriteLine ("Finally block executes");
}
```

When used with *catch* blocks, the *finally* block must be at the end of the sequence. It will execute regardless of whether your program throws or catches an exception, as shown in the following snippet:

```
try
{
    main.CauseProblem ();
}
catch (FormatException e)
{
    Console.WriteLine ("Format Exception");
}
finally
{
    Console.WriteLine ("Finally block executes");
}
```

Because the *finally* block is guaranteed to execute, it is useful for cleaning up and freeing any resources that you might have allocated in the *try* block. You should be careful, however, to make sure your program does not later attempt to use any resources that were freed in the *finally* block.

WHAT YOU MUST KNOW

At some point in your C# programming experience, your program is going to throw an exception. Using C# exception-handling statements and techniques, your program can respond to and correct serious and unexpected errors. In this lesson, you learned about the *try* statement, in which you write a guarded block of code. If your program throws an exception in a *try* block, control passes immediately to a *catch* block. You learned how to write *catch* blocks to handle exceptions of different types, and how to derive your own exception class from the generic C# *Exception* class. You also learned how to add a *finally* block to your exception-handling sequence. In Lesson 16, "Using Array Variables to Store Multiple Values," you will learn how to declare and use arrays in C#, and how to use exceptions to catch errors in using arrays. Before you continue with Lesson 16, if you disabled Just-In-Time Debugging for this lesson, use the steps in "Throwing an Exception" to re-enable it. Then, make sure you have learned the following key concepts:

- The *try* block contains "guarded" code in which your program may detect exceptions generated by your program or the Common Language Runtime code.

- An unhandled exception will cause your program to exit. You can "catch" an exception and write code to handle the error in a *catch* block.

- When C# or your code throws an exception, it must use an object derived from the *Exception* class.

- When your program throws an exception, it looks for a *catch* block that uses an object identical to the object that was thrown.

- Your program may use multiple *catch* blocks to handle exceptions of different types.

- The *finally* block is guaranteed to execute. When you use a *finally* block with *catch* blocks, it must be the last block in the chain.

USING ARRAY VARIABLES TO STORE MULTIPLE VALUES

*a*s they execute, programs store values in memory locations identified by variables. In previous lessons, your programs have used one variable to store one value. In Lesson 14, "Using Structures In C# to Group Related Data," you used a structure as a *container* to hold multiple variables, including variables of different types. You will often find it convenient to store multiple values in a single variable, such as the grades of all the students in a class, or the hourly temperatures for a day. To do this, you use an *array*. In C#, an array is a data mechanism that holds multiple values in *elements*. You then access these elements using an *index* value. In this lesson, you will learn how to declare and use arrays in a C# program. By the time you finish this lesson, you should understand the following key concepts:

- Using an array, your programs can store multiple values in a single variable. All the values use the same variable name.
- The C# array is a reference-type value derived from the *System.Array* class.
- Each value in an array is an element. You access elements using an index into the array.
- To access the elements in an array, you must identify the elements by the name of the array and an index. Visual C# *always* uses 0 to identify the first array element.
- If you try to access an array element using an index that is too large for the size of the array, C# will throw an exception.

Arrays Let Your Program Store Multiple Values in a Single Variable

Variables represent memory addresses where your program stores values. The amount of memory reserved to hold a value is just large enough to hold a value of the variable's data type. An *int* type uses four bytes, a *short* type uses two bytes, and so on. One variable is usually all you will need to store a value.

However, many times your programs will deal with several related values such as today's 24 hourly temperatures at your local airport. If you had to declare individual variables for each of these values, your code quickly would become unwieldy. You would need *Temp01* for the 1 a.m. temperature, *Temp02* for the 2 a.m. temperature and on up to *Temp24* for the midnight temperature. You would have to access each of these values individually because you would not be able to use *loops* to access the values sequentially. This could be a very tedious process.

To make this process easier, Visual C# lets you declare related variables in an *array*. An array is a data structure that allows a single variable to hold more than one value. (A "data structure" is a way of organizing information to make it easier to perform operations on the information, or data).

In C#, arrays are reference-type variables, and you derive arrays from the *System.Array* class. You do not need to declare a specific class to use arrays. When you declare an array, the C# compiler does this for you.

Although you may declare an array of any data type, all the elements of an array must be the same data type, which is the *element type* of the array. In other words, you cannot store values of type int and type float within the same array.

When you declare an array variable, your program sets aside enough memory to hold all the elements of the array. For example, you learned earlier that Windows needs four bytes to store a value of type *int*. If you declare an array large enough to hold 10 values of type *int*, your program will allocate 40 bytes of memory for the values. In C#, the number of bytes allocated actually will be more than the 40 bytes needed for the values, because at the same time, you are declaring an instance of the *System.Array* class.

Declaring Arrays in Your C# Program

Because an array in C# is a reference-type variable, you must use the *new* operator when you create the array. You can declare the array in one statement and create the array elements in another statement, or you can declare and create the array in a single statement. If you have studied C or C++, the syntax is slightly different.

To declare an array, you first write the element type of the array. This may be a fundamental data type such as *int*, *char*, or *long*, or it may be a user-defined data type, such as the name of a class or a structure. After the element type, you write an *empty* set of brackets. The C# compiler uses the empty brackets as a signal that the declaration is for an array and that you intend to declare an instance of *System.Array*. (Technically, *System.Array* is an *abstract* class and you cannot declare an instance of it directly. However, when you declare an array, the C# compiler derives a new class from *System.Array*. You do not have to define the class explicitly.) Immediately after the brackets, you type the name that you want to use for the array variable. The following line shows a complete declaration of an array of type *int*:

```
int [] MyInts;
```

The preceding statement does not actually *create* the array. It simply declares an instance of an array type class. At this point, neither the C# compiler nor your program have any idea how many elements the array will contain. If you declare an array in this way, you must use a second statement to create the array. You do this with the *new* operator, specifying once again the element type and a second set of brackets. This time, however, you must give the array a size. The following snippet declares an array variable, then creates an array containing 24 elements:

```
int [] Temps;
Temps = new int [24];
```

You can simplify this step by combining the declaration and assignment in a single statement. Normally, this is how you will declare arrays, as in the following line:

```
int [] Temps = new int [24];
```

After you create the array, the size is fixed. You cannot add extra elements to it. This does not mean that you may not *recreate* the array with a new size, but you should remember that you will lose the old values that you stored in the newly created array, as shown in the following snippet:

```
int [] Temps = new int [24];
// Some statements that use Temps array
Temps = new int [48];      // Values stored in array are lost
```

In the preceding snippet, you reserved a new block of memory for the values. Eventually the Common Language Runtime garbage collection will remove the old block, leaving only the new array. To save the old values, you would have to create a new array variable, then assign the old values from one array to the other in a *loop*. You will learn about loops in Lesson 21, "Repeating Statements Within a C# Program."

You can initialize the elements of an array when you create the array. You write the values inside a set of curly braces immediately after the brackets that contain the size of the array:

```
int [] Temps = new int [12] {68, 72, 56, 90, 75, 65,
                             82, 73, 87, 64, 88, 79};
```

You must be careful to provide exactly enough values to initialize every element in the array. If you specify too many or too few, the compiler will issue an error and will not compile your program. (This is a marked departure from C++ syntax,

in which you can specify fewer initializers than elements. When you do this, C++ assigns the remaining elements the last value in the initializer list.)

Letting the Compiler Determine the Size of an Array

If you specify the size of an array when you create and initialize the array, you must make sure the number of initializing values matches the number of elements in the array. This means you have to count the values, and if you later change the size of the array, you will have to change the size value. You can avoid having to do this by letting the compiler calculate and set the size of the array.

*Write the variable declaration as you normally would, but **without** a size inside the brackets; then provide the list of initializer values after the empty brackets, as shown below:*

```
int [] Temps = new int []{68, 72, 56, 90, 75, 65,
                          82, 73, 87, 64, 88, 79};
```

When you declare an array variable this way, the compiler will count the number of values between the braces, then set aside enough memory to hold all the values.

*Your program still must know how many elements are in the array. You can get this value using the **System.Array**'s built-in **GetCount()** method. This function takes the **dimension** of the array as a parameter and returns the number of elements in the array. Later in this lesson you will learn to declare and use multidimensional arrays. The arrays that you have declared and created so far are one-dimension arrays, and the dimension is always 0. The following statement gets the size of a one-dimension array:*

```
int size = Temps.GetCount(0);
```

*The **GetCount()** function returns the value of the **Length** property, which is a **public** read-only property that you can access directly. You could get the same value as the above statement using the following syntax:*

```
int size = Temps.Length;
```

*For a multidimensional array, **Length** is the count of all of the elements in all of the dimensions of the array. Because **Length** is a read-only property, trying to change the value will result in a compile error. The value is set when you create the array giving the **System.Array** class the number of elements in the array.*

Understanding Array Types

An array is a reference-type variable and is a class instance derived from the *System.Array* class. Thus, arrays are actually *object* types rather than simple variable types. Because an array is an object, it has methods, fields and properties in the *System.Array* class that you can use to manipulate the array.

When you create an array, you create an *array type* by declaring the element type, the number of dimensions and the number of elements in the array. The C# compiler uses these characteristics to create an array type automatically; you do not have to add any statements to create the array type.

Because an array is an instance of a class, you may use the methods and properties in the *System.Array* class to access and manipulate the array. Table 16.1 lists some of the useful *static public* methods that are members of *System.Array*.

Method	Purpose
BinarySearch()	Performs a binary search for a value in a one-dimension array. The array must be sorted.
Clear()	Sets a range of elements in an array to 0 or NULL.
Copy()	Copies the values in one array to another array. If the arrays are of different data types, this method will perform any *downcasting* (such as converting float to int) needed.
IndexOf()	Returns the index of the *first* occurrence of a value in a one-dimensional array.
LastIndexOf()	Returns the index of the *last* occurrence of a value in a one-dimensional array.
Reverse()	Reverses the order of the elements in a one-dimension array.
Sort()	Sorts the elements in a one-dimensional array. This method places the smallest value at index 0.

Table 16.1. Static member methods in the System.Array class.

The methods Table 16.1 lists are *static* methods, so you access them using the *Array* class rather than the variable name of the instance. For example, you call the *Reverse()* method using *Array.Reverse()*. The following short program, *StatArry.cs*, uses these methods. Notice that the temperature 75 appears twice in the array.

```
namespace nsStaticMethods
{
    using System;
    class clsMain
    {
        public static void Main ()
        {
            int [] Temps = new int []{68,72,56,90,75,65,
                                82,73,87,75,88,79};
// Show the original array elements
            Console.Write ("The temperatures are ");
            foreach (int x in Temps)
            {
                Console.Write (x + " ");
            }
            ShowArray (Temps);
// Copy the array
            int [] Sorted = new int [Temps.Length];
            Array.Copy (Temps, Sorted, Temps.Length);
// Sort the array
            Array.Sort (Sorted);
            Console.Write ("\nThe sorted elements are ");
            foreach (int x in Sorted)
            {
                Console.Write (x + " ");
            }
            ShowArray (Sorted);
// Do a binary search for 82
            int Index = Array.BinarySearch (Sorted, 82);
            if (Index < 0)
            {
                Console.WriteLine ("Temperature 82 was " +
                                "not found");
            }
            else
            {
                Console.WriteLine ("\nTemperature 82 is " +
```

```
                                        "at index " +
                                    Array.BinarySearch (Sorted, 82));
            }
// Reverse the elements
            Array.Reverse (Sorted);
            Console.Write ("\nThe sorted array elements " +
                            "in reverse order are \n");
            foreach (int x in Sorted)
            {
                Console.Write (x + " ");
            }
            ShowArray (Sorted);
// Do a binary search for 82
            Index = Array.BinarySearch (Sorted, 82);
            if (Index < 0)
            {
                Console.WriteLine ("Temperature 82 was " +
                                    "not found");
            }
            else
            {
                Console.WriteLine ("\nTemperature 82 is " +
                        "index " +
                            Array.BinarySearch (Sorted, 82));
            }
        }

        static void ShowArray (int [] Temps)
        {
            Console.WriteLine ();
// Find the first occurrence of 56
            Console.WriteLine("The first occurrence of 75 " +
                            "is at index " +
                            Array.IndexOf (Temps, 75));
// Find the last occurrence of 56
            Console.WriteLine("The last occurrence of 75 " +
                            "is at index " +
                            Array.LastIndexOf(Temps, 75));

        }
    }
}
```

When you compile and run StatArry.cs, you should see the following output. Notice that after you reverse the sort in the array, the binary search for 82 fails. The binary search is designed for a sorted array beginning with the smallest value in the first element.

```
The temperatures are 68 72 56 90 75 65 82 73 87 75 88 79
The first occurrence of 75 is at index 4
The last occurrence of 75 is at index 9

The sorted elements are 56 65 68 72 73 75 75 79 82 87 88 90
The first occurrence of 75 is at index 5
The last occurrence of 75 is at index 6

Temperature 82 is at index 8

The sorted array elements in reverse order are
90 88 87 82 79 75 75 73 72 68 65 56
The first occurrence of 75 is at index 5
The last occurrence of 75 is at index 6
Temperature 82 was not found
```

Accessing Array Elements

For indexing purposes, C# arrays always begin at element 0. There is no option to set the first element to an index of 1 as there is in Visual Basic. Thus, the index of the last element in an array will be the number of elements minus 1. If you have 10 elements in an array, the index of the last element is 9. (VB programmers should note that Visual Basic.NET does not support the Option Base statement.)

To access an array element, you write the name of the array variable followed by a set of brackets containing the element index, as in the following snippet:

```
int [] MyArray = new int [10];
MyArray[4] = 42
```

In this code, the *4* is the index into the array and indicates you want to access the fifth element in the array.

Unlike C and C++, C# performs *bounds checking* on the index before allowing you to access the array element. If you try to access an element using an index that is not a part of the array, your program will throw an exception. If you do not catch and handle the exception, your program will terminate.

Unless you are using one of the C# safe methods of accessing array elements, such as an *Array* class member method or the *foreach* loop statement, it is a good idea to protect critical parts of your code in *try* and *catch* blocks.

The following program, *ArrExcep.cs*, shows a common error: getting the length of the array, and then trying use that value as an index into the array. The code shows how to use *try* and *catch* blocks, and also shows what happens when you try to access an array element that does not exist:

```csharp
namespace nsArrayException
{
    using System;
    class clsMain
    {
        public static void Main()
        {
            int [] Temps = new int []{68,72,56,90,75,65,
                                      82,73,87,75,88,79};
            int index = Temps.Length;
            try
            {
                Console.WriteLine ("The last temperature " +
                                   "in the array is " +
                                   Temps[index]);
            }
            catch (IndexOutOfRangeException e)
            {
                Console.Write (e.Message + "\n");
                Console.Write ("The index value {0} " +
                               "is invalid. The maximum " +
                               "index allowed is {1}",
```

```
                    index, Temps.Length - 1);
        }
      }
    }
}
```

When the program tries to access an array element using an index that is *out of bounds*, the program throws an exception. After you catch the exception within your program, your code can make whatever corrections are necessary. In this case, the program prints an error message.

Declaring Arrays of Classes and Structures

You may declare and create arrays of reference data types, but simply declaring and creating the array does not create the objects. The array itself is a reference object, and when used for reference-type values, creating the array only sets aside space to hold reference to the objects. You still must create the object and assign references to the array elements.

This is a sharp departure from the way C++ handles arrays of objects. In C++, when you create the array, you create the objects at the same time. If you have studied C++, it is important that you remember this difference. The following short program, *ClsArr.cs*, will cause your C# program to throw an exception, although the general syntax is correct for C++:

```
namespace nsClassArray
{
    using System;
    class clsTestClass
    {
        public int x = 42;
    }
    class clsMain
    {
        public static void Main()
        {
            clsTestClass [] test = new clsTestClass [10];
```

```
        try
        {
            Console.WriteLine ("x in test[4] = {0}",
                                test[4].x);
        }
        catch (NullReferenceException e)
        {
            Console.WriteLine (e.Message);
        }
    }
  }
}
```

The exception message will tell you that you have attempted to "dereference a null object reference." This may sound cryptic, but what it means is that when you created the array, the elements were assigned a value of 0, or *null*. There are no objects to reference yet.

There are several ways you can create the objects and place them in the array elements. First, you could create them in an initializer list, just as you initialized an array of type *int* when you create the array. The following snippet shows how:

```
clsTestClass [] test = new clsTestClass []
                    {
                    new clsTestClass(), new clsTestClass(),
                    new clsTestClass(), new clsTestClass();
                    };
```

This works for a small array, but if the number of class objects is much more than about 10 or 12, it gets a bit unwieldy. It really is not much easier than creating the objects in individual statements. The advantages are that you do not need to specify the size of the array, and you can pass virtually random values through the constructors if necessary.

A second method is to create the object in a loop. You can use the *Length* property to control the loop. The following uses a *for* loop to create the 12 objects that will be elements of an array.

```
clsTestClass [] test = new clsTestClass [12];
for (int x = 0; x < test.Length; ++x)
{
    test[x] = new clsTestClass();
}
```

This method is clean and fast. If you add more elements to the array later, you do not have to add more initialization statements. However, if you must pass parameters to a class constructor, they will have to be computed in the loop or held in a separate array.

Another method is to provide a *static* method member that will initialize the members of the class instance, and at the same time create a copy of the class itself. The method should return a reference to a class object. The following program, *InitArr.cs*, uses a *static* method named *Initialize()* to declare an instance of the class and return it to the calling statement, which saves the return value in the array:

```
namespace nsClassArray
{
    using System;
    class clsTestClass
    {
        public static clsTestClass Initialize ()
        {
            clsTestClass obj = new clsTestClass();
// Initialization code here
            return (obj);
        }
        public int x = 42;
    }
    class clsMain
    {
        public static void Main()
        {
            clsTestClass [] test = new clsTestClass [10];
            for (int x = 0; x < test.Length; ++x)
```

```
    {
        test[x] = clsTestClass.Initialize();
    }
    try
    {
        Console.WriteLine ("x in test[4] = {0}",
                            test[4].x);
    }
    catch (NullReferenceException e)
    {
        Console.WriteLine (e.Message);
    }
    }
  }
}
```

You cannot use a constructor for this purpose because a constructor does not have a return type. Remember, however, that static methods may only access static fields and properties, and other static methods. Also, because it is a static member, you can test for certain conditions before creating an instance of the class.

Using a for Loop to Repeat Statements

Very often in programming, you must repeat one or more statements several times, as you did in this section to create the reference-type variables and assign them to the array elements. When you do this, you are causing your program to loop.

C# contains several forms of loops, but one of the most common is the for loop. The for loop consists of four parts. The first three parts are the loop control statements, which you write inside a set of parentheses after the for keyword. You separate these three statements with semicolons, as in the following example. Notice that there is no semicolon at the end of the for statement:

```
for (x = 0; x < 10; ++x)
{
    // Loop statements that will repeat
}
```

*The first statement in the loop control is the initializer. The **for** statement performs the initializer only once, when the program control first enters the loop. You can declare variables in the initializer. End the initializer statement with a semicolon.*

*The second statement in the loop control is the loop condition. Your program will execute this expression each time it performs the loop, including the first **iteration** of the loop. If the statement is true, the loop will execute; if it is false, the loop will terminate without executing.*

*The third statement of the loop control is the **increment**. Your program executes this expression at the end of each loop iteration. Despite its name, you can use any expression in the increment. Usually, though, you will use this expression to increment or decrement a **control variable**.*

*The fourth part of the **for** loop is the statement or block of statements that your program will execute each time it performs the loop. This may be a simple statement, but more often than not it will be a **compound** statement that you write between a set of braces.*

Using Strings In C#

Unlike C and C++, C# does define a *string* data type. The *string* type is not a fundamental data type, however. It is convenient to think of *string* as a class derived from *System.Array*. As a class, it declares an array of characters and makes available all the methods and properties of the *System.Array* class.

The keyword *string* actually is an alias for the *System.String* class. To make it work like a fundamental data type, the class redefines how operators such as the assignment operator, the addition operator, and the inequality operators function through *overloading*. You will learn more about overloading in Lesson 32, "Overloading Functions and Operators."

The characters in a *string* are *Unicode* characters. In the Unicode scheme, a single character is 16 bits rather than the eight bits in a *char*. A Unicode character, then, can be any one of up to 65,537 characters, considerably more than the 256

available in the single-byte *char* type in C and C++. In C#, even the *char* fundamental data type is a 16-bit character.

As an array, a *string* is immutable after you declare an object and assign it a value. Although the class contains methods that allow you to modify the string, they actually create new instances of the *String* class.

Declaring and Using Multidimensional Arrays

Arrays are reference-type variables that let your program store multiple values using only a single variable name. You access the array elements to set or retrieve the values using an *index* into the array. Up to this point, the arrays you have dealt with have had only one dimension; they have had only a single relationship.

Sometimes you will need to describe arrays in more than one dimension. For example, you might want to store the 24 hourly temperature for all the days in a month. In this case, you would declare a two-dimensional array. One dimension would contain the days of the month, and the second dimension the temperatures for a particular day. You access the elements using two indexes, one for the day and another for the temperature.

To declare a multidimensional array, write the declaration using a comma for each of the additional dimensions you want in the array, then specify the size of the array using the new operator. The following statement declares a three-dimensional array with 10 elements in the first dimension, 15 elements in the second, and 20 elements in the third:

```
int [,,] MultiArr = new int[10,15,20];
```

To access the elements in a multidimensional array, you access it using a single set of square brackets, specifying the element in each dimension that you want to access. The following would get the value of the third element in the first dimension (remember the first element is element 0), the eighth element in the second dimension, and the sixth element in the third:

```
int x = MultiArr [2, 7, 5];
```

C# also allows you to declare "jagged" arrays. Remember from the section on "Declaring Arrays of Classes and Structures" that when you declare an array of a reference data type, you do not actually create the reference objects in the declaration. An array is a reference type, so when you declare a two-dimensional array, you are declaring an array of arrays.

To declare a two-dimensional jagged array, you write the data type followed by two sets of brackets, then the variable name, as shown in the following example:

```
int [][] MonthTemps;
```

You can initialize the variable at the same time. In this case, you provide the size of the first array dimension, but you must omit the size of the second dimension:

```
int [][] MonthTemps = new int [31][];
```

To complete the array, you then would have to create the arrays for the second dimension and assign them to the elements in the first dimension:

```
int [][] MonthTemps = new int [31][];
for (int x = 0; x < MonthTemps.Length; ++x)
{
    MonthTemps[x] = new int [24];
}
```

Using this system, you would store the temperature for 2 p.m. (1400 hours; midnight is element 0) on the 16th day of the month by using two indexes:

```
MonthTemps[15][14] = 78;
```

This is considerably different from arrays in C and C++, but there are some advantages to the C# method. For example, in C and C++, the number of elements in all the arrays in the same dimension must be the same, whether you use all the elements or not. This often leads to wasted memory. In C#, the lengths do not have to be the same, leading to the term "jagged array."

This is easily visible if you extend the temperature array to three dimensions. You want to store all the temperature for an entire year so that you can index them by month, day, and hour. The longest month is 31 days, so in C and C++ you would have to declare every element in the month dimension to be 31. You will use that many days only in seven of the months, and the other elements would go unused.

In C#, you can declare a jagged array in the month dimension, using only the memory you need. To do this, you declare the three-dimension array using three sets of brackets after the data type. You must specify only the size of the first dimension, and the second and third dimensions must be empty, as shown in the following example:

```
int [][][] YearlyTemps = new int [12][][];
```

It is easier to visualize this in a program. The following program, *YearTmps.cs*, creates such a three-dimension jagged array. The lengths of the days of the month will be stored in a separate array to make it easier to use them in a loop. This program will not even try to initialize the 365 elements in the array (it will assume the year is not a leap year, so there are only 28 days in February). Remember, too, that the arrays are zero-base indexed, so January is month 0 and December is month 11, and the first day of the month is 0. For the hours dimension, this code will assume the 0 element is 0 hours, or midnight.

```
namespace nsYearly
{
    using System;
    class clsMain
    {
        static int [] DaysInMonth = new int [12]
                              {31, 28, 31, 30, 31, 30,
                               31, 31, 30, 31, 30, 31};
        public static void Main ()
        {
            int [][][] YearlyTemps = new int [12][][];
            for (int x = 0; x < YearlyTemps.Length; ++x)
            {
                YearlyTemps[x] = new int [DaysInMonth[x]][];
                for (int y = 0;
                     y < YearlyTemps[x].Length;
                     ++y)
                {
                    YearlyTemps[x][y] = new int [24];
                }
            }
//
// Add code here to initialize the array.
// Here we will initialize only one element
//
            YearlyTemps[5][15][14] = 68;

            Console.WriteLine ("The temperature for 2 " +
                             "p.m.June 16 was " +
                             YearlyTemps[5][15][14]);

        }
    }
}
```

To prove that the array is jagged, try assigning values to an hour the 29th of each month (the 28th element). The program will throw an exception when you attempt to store a value in YearlyTemps[1][28][any hour].

Fortunately, you will not need three-dimension or greater arrays very often. You should know how to create them, and how to use jagged arrays, when the occasion arises, however.

WHAT YOU MUST KNOW

In this lesson you learned how to declare and use arrays in C#. You learned how to initialize an array at the same time you declare it in your program, and how to use many of the *System.Array* member methods. You also tackled a somewhat difficult subject in C#, using multidimensional arrays. Arrays of more than one dimension are much simpler in C or C++, but the C# method provides for more flexibility and allows the use of "jagged" arrays. Lesson 17, "Using Increment and Decrement Operators," will be somewhat easier. You will learn how to use the increment and decrement operators to step the value of a variable up or down by one. Before you continue with Lesson 17, however, make sure you have learned the following key concepts:

- Arrays are a data mechanism that allows your program to store multiple values of the same data type using a single variable name.

- Arrays in C# are reference-type variables. You must create them using the *new* operator.

- You access the elements of an array using an index value. In C#, the first element of an array always has an index of 0.

- You can initialize the elements of an array when you create the array by writing the initial values inside a set of braces after the *new* operator.

- When you create an array of reference-type objects, the objects themselves are not created by the declaration. You must create the object and assign a reference to the array element.

- C# builds arrays of more than one dimension by creating arrays of reference objects. In a multidimensional array, your declaration can specify only the size of the first dimension.

- The individual arrays in a multidimensional array do not have to be the same length. This permits the use of "jagged" arrays.

namespace nsIncrement
{
    using System;

    class Main

    static public void Main ()

    val = 5;
    Console.WriteLine ("The value of val before " +
    "the increment is " + val);

    int sum = ++val

    = ++val + 2;
    Console.WriteLine ("The value of val after " +
    "the increment is " + val);

    Console.WriteLine ("The sum is " + sum);

}

USING INCREMENT AND DECREMENT OPERATORS

*i*n Lesson 16, "Using Array Variables to Store Multiple Values," you learned how to create and use array variables in a C# program. To access the elements of an array, you wrote the index of the element you wanted to access inside a set of brackets after the variable name. Very often in accessing the elements of an array, you will need to add 1 to the index to get to the next element, or subtract 1 from the index to get the previous element. When you use a variable to hold the index value, you can perform the increment and decrement operations by adding to or subtracting from the variable's value. The process of stepping a variable's value up or down by one is so common that C# provides special operators just to perform this operation. In this lesson, you will learn about the increment and decrement operators and how to use them. By the time you finish this lesson, you will understand the following key concepts:

- When you add a fixed amount to the value of a variable, you are *incrementing* the variable. You can use increments to step the value of the variable up until it reaches a certain value.

- Decrementing the value of a variable is the process of subtracting a fixed amount from the value of a variable.

- Incrementing and decrementing occur often in programming and are key operations in constructing loops.

315

- C# provides an increment operator—a set of two plus signs—to increment a variable by one. The decrement operator—a set of two minus signs—decreases the value of a variable by one.

Incrementing and Decrementing Variables

Throughout your C# programming, you will find that you often must increase or decrease the value of a variable to keep track of a particular operation. You may need to step through the elements of an array by adding 1 to an index variable. You may need to perform an operation several times, such as printing a number of files, which might involve subtracting 1 from the value of a *control* variable until it reaches 9.

The process of adding a fixed amount to the value of a variable is called *incrementing*. When you subtract a fixed amount from the value of a variable, you are *decrementing* the value. The most common fixed value in increment or decrement operation is 1. You could perform these operations by writing statements that add or subtract 1 from a variable and then assign the result to the variable itself, as in the following statements:

```
iVar = iVar + 1;     // Increment the value
iVar = iVar - 1;     // Decrement the value
```

The process of increasing or decreasing the value of a variable by 1 is so common in programming that C# provides special operators just for this purpose. Using the *increment* operator, ++, on a variable is the same as adding 1 to the value of the variable. Using the *decrement* operator, --, is the same as adding 1 to the variable's value. In C#, you could write the preceding two lines as follows:

```
++iVar;     // Increment the value
--iVar;     // Decrement the value
```

There are a few things you should remember about the increment and decrement operators. First, you must type them without any spaces between the two plus signs or the two minus signs. Second, you can only use them on value-type variables or properties; you cannot use them on reference-type variables or constant values such as a numbers.

You can write the operators before or after the variable, but the difference affects the result of the operation when you use the operators in an expression.

Using Prefix Expressions

If you write an increment or decrement operator before the name of a variable in an expression, the operation is a *prefix* operation. Your program will increment or decrement the value of the variable and use the new value in the expression. This will affect the result of an expression.

In the following short program, *PreOp.cs*, the result of the addition is 8 because the increment operation is performed before the addition operation:

```
namespace nsIncrement
{
    using System;
    class clsMain
    {
        static public void Main ()
        {
            int val = 5;
            Console.WriteLine ("The value of val before " +
                                "the increment is " + val);
            int sum = ++val + 2;
            Console.WriteLine ("The value of val after " +
                                "the increment is " + val);
            Console.WriteLine ("The sum is " + sum);
        }
    }
}
```

First, you assign 5 to the value of *val* and print the result to the screen. In the next statement, the increment operator appears before the variable name, so your program performs the increment before using the value. The result of the increment is 6, which is used in the addition operation, resulting in a value of 8.

When you compile and run this program, you should see the following output:

```
The value of val before the increment is 5
The value of val after the increment is 6
The sum is 8
```

The same is true of the decrement operator. If you substitute *−−val* for *++val* in the addition statement, the value of *val* before the addition is performed will be 4, resulting in a sum of 6.

Using Postfix Expressions

In a *postfix* operation, you write the increment or decrement operator after the variable name. When written this way, your program will use the current value in the expression, then increment or decrement the value.

The following short program, *PostOp.cs*, performs the same operations a *PreOp.cs*, but uses the increment operator in a postfix operation.

```csharp
namespace nsIncrement
{
    using System;
    class clsMain
    {
        static public void Main ()
        {
            int val = 5;
            Console.WriteLine ("The value of val before " +
                                "the increment is " + val);
            int sum = val++ + 2;
            Console.WriteLine ("The value of val after " +
                                "the increment is " + val);
            Console.WriteLine ("The sum is " + sum);
        }
    }
}
```

When you compile and run *PostOp.cs*, you should see the following output. Compare this with the output you saw when you ran *PreOp.cs*.

```
The value of val before the increment is 5
The value of val after the increment is 6
The sum is 7
```

The only difference between the two programs is that *PreOp.cs* used the increment operator in a prefix operation, and *PostOp.cs* used it in a postfix operation. The position, however, affected the outcome of the addition operation and thus the value of *sum*.

Use Caution in Writing Increment and Decrement Operators

*Whether you use increment and decrement operators in prefix or postfix operation, you must remember to type the operators without any spaces between the two plus or minus signs. If you type a space between the symbols, the C# compiler will interpret them as two **unary** operators, and the increment or decrement operations will not be performed.*

*In the case of prefix operations, you should be aware that the compiler will **not** issue an error if you accidentally place a space between the symbols. To show this, return to the **PreOp.cs** program in the previous topic and add a space between the two so the statement looks like the following:*

```
int sum = + + val + 2;    // Add a space
```

*Recompile and run the program. You will not get an error, but the increment operation will not occur. The compiler first sees a unary plus sign, then another unary plus sign, so the value of **val** remains unchanged.*

In the case of a postfix operation, the compiler may or may not issue an error, depending on the context of the statement. If the increment or decrement is the last expression in a statement, or if it is a standalone statement as in the

following example, the compiler will issue an error because it expects an operand after the last plus sign:

```
val+ +;
```

If, however, the increment or decrement occurs at the beginning or in the middle of a statement, the compiler again will recognize the symbols as unary operators:

```
int sum = val+ + + 2;    // Add a space
```

In the above statement, the compiler sees the first plus sign as an addition operator. The next two plus signs are unary plus operators that the compiler applies to the constant 2. The increment never takes place.

WHAT YOU MUST KNOW

In this lesson, you learned how to use the increment and decrement operators. C# provides the operators because the process of incrementing or decrementing a variable's value by 1 is a common operation in programming. You learned how to use these operators in prefix and postfix operations, and how each affect the outcome of an expression. In Lesson 18, "Writing Expressions in C#," you will learn how C# evaluates expressions and how it uses operators in expressions. Understanding expressions will help you to avoid a number of common programming errors. Before you continue with Lesson 18, however, make sure you have learned the following key concepts:

- ❎ Incrementing a variable is the process of adding a fixed value such as 1 to the value of the variable. C# provides a special operator, ++, for this operation.

- ❎ Decrementing a variable is the process of subtracting a fixed value such as 1 from the value of the variable. The C# decrement operator, --, performs this operation.

- You can use the increment or decrement operators in a *prefix* operation by writing the operator before the variable name. When you do this, your program performs the increment or decrement operation before using the value of the variable.

- In a *postfix* operation, you write the operator after the name of the variable. In this case, your program uses the value of the variable before performing the increment or decrement operation.

- The increment and decrement operators contain more than one symbol, and you must remember to type the symbols without any spaces between them.

argument to an integer. A non-number
FormatException
Celsius;

Celsius = args[0].ToInt32();
}
catch (FormatException)
{
Console.WriteLine ("please enter a " +

WRITING EXPRESSIONS IN C#

expressions in C# are combinations of operators, values and variables that return a result or cause something to happen, such as a function call. C# statements contain one or more valid expressions. Understanding expressions and how your program evaluates expressions will help you to avoid a number of common programming errors. In this lesson, you will learn about C# expressions and operators and how to use them. By the time you finish this lesson, you will understand the following key concepts:

- An expression is a group of operators and identifiers that equate to a value or generate an effect in Visual C#.
- Expressions may contain other subordinate expressions that your program may have to evaluate first.
- C# assigns a *precedence* to each operator to assure that programs evaluate expressions consistently and in an orderly manner.
- You can use parentheses to group operators and identifiers and to change the order in which C# evaluates them.
- Boolean expressions are logical expressions that equate to *true* or *false*.

How C# Evaluates Expressions

When you combine operators, variables and constant values in a programming statement, you are writing an *expression*. In programming, an expression is a group of one or more tokens that evaluate to a value. An expression may be as

simple as a constant value such as the number *42*, or it may be a complex formula such as *9 * c / 5 + 32* that converts a temperature from degrees Celsius to degrees Fahrenheit.

When you use an expression in a program line to equate a value, the result is a statement. Unlike C and C++, the C# language requires that a statement does something useful. For example, the following statement is valid in C and C++, but will produce an error in C#:

```
42;
```

Your program would evaluate the expression *42* and immediately discard the result because you do not use the resulting value. C# requires that you use the result of an expression to assign the value to a variable or to call a function. Expressions that increment or decrement a variable or use the *new* operator to create a reference-type object also can be complete statements. This leads to the curious result that you can create an object in a manner such that it cannot be used, as in the following:

```
new int [10];
```

This is a valid statement in C#, but you cannot use the result of the expression because you have no way of identifying the array. There are theoretical reasons why this must be a valid statement, but you should be aware that you can write such a statement and get a compiler error. Ordinarily you will want to capture the result of such an expression in an assignment statement:

```
int [] MyInts = new int [10];
```

The variables, operators and values on the left side of the equals sign make up a complete expression, as in the following assignment statement:

```
int Fahr = 9 * Celsius / 5 + 32;
```

In such an assignment statement, your program will evaluate the expression on the right side of the equals sign fully before it attempts to assign the result to the variable on the left side.

The following command line program, *C2F.cs*, uses expressions to convert degrees Celsius that you enter on the command line to degrees Fahrenheit and display the result:

```csharp
namespace nsTemps
{
    using System;
    class clsMain
      {
        static public void Main (string [] args)
        {
            if (args.Length == 0)
            {
                Console.WriteLine ("Please enter a " +
                                "temperature in Celsius");
                return;
            }
//  Convert the argument to an integer. A non-number
//  will throw a FormatException
            int Celsius;
            try
            {
                Celsius = Convert.ToInt32(args[0]);
            }
            catch (FormatException)
            {
                Console.WriteLine ("Please enter a " +
                                "numeric value");
                return;
            }
            int Fahr = 9 * Celsius / 5 + 32;
            Console.WriteLine ("{0} degrees Celsius = " +
```

```
                              "{1} degrees Fahrenheit",
                              Celsius, Fahr);
        }
    }
}
```

C2F.cpp uses a *command line argument*. After you compile the program, run it by typing a number after the program name as shown below:

```
C2F 20 <Enter>
```

The program will respond by displaying the following output on your screen:

```
20 degrees Celsius = 68 degrees Fahrenheit
```

Understanding Operator Precedence

When you studied algebra and other fields of mathematics, you learned that you had to perform certain operations in a specific order to assure that formulas generate consistent results. For example, you learned that when you combine multiplication and addition, you must perform the multiplication first, then the addition, as in the following example:

```
4 + 7 * 6 = 46
```

Here you evaluate *7 * 6* and get the result, *42*, before you perform the addition. If you performed the addition first, the result would be very different.

C# also treats different operators with a different priority, called the *operator precedence*. It is necessary to evaluate some expressions before others. For example, when you add a value to an element in an array, you first must determine the value of the array element, which means the array subscript operator, [], must have a higher precedence than the addition operator.

Table 18.1 lists the C# operators and their precedence. Notice that the 11 assignment operators have the lowest precedence and thus are at the bottom of the list. This assures that your program will evaluate expressions on the right side of an assignment expression before making the assignment.

Operator	Meaning	Example
.	Member of	object.member_name
[]	Subscript	array[element]
()	Function call	FunctionName (arguments)
()	Sub expression	(Expression)
++	Postfix increment	variable++
−−	Postfix decrement	variable−−
new	Allocate memory	new type
typeof	Type retrieval	typeof(object type)
sizeof	Size of type	sizeof(object type)
checked	Range checking	checked (expression)
unchecked	Range checking	unchecked (expression)
!	Logical not	!variable
++	Prefix increment	++variable
−−	Prefix decrement	−−variable
~	Ones complement	~variable
+	Unary plus	+42 or +variable
-	Unary minus	-42 or −variable
()	Cast	(type) variable
*	Multiply	expression * expression
/	Divide	expression / expression
%	Modulo	expression % expression
+	Addition	expression + expression
-	Subtraction	expression − expression
<< and >>	Shift left, Shift right	a << b or a >> b
<	Less than	a < b
>	Greater than	a > b
<=	Less than or equals	a <= b
>=	Greater than or equals	a >= b
is	Compatibility	expression *is* object type
== or !=	Equality	a == b or a != b
&	Bitwise AND	a & b
|	Bitwise OR	a | b

Operator	Meaning	Example (cont.)
^	Bitwise XOR	a ^ b
&&	Conditional AND	expression && expression
||	Conditional OR	expression || expression
?:	Conditional	x = a ? b : c
	Assignment	= *= /= %= += -= <<= >>= &= ^= |=

Table 18.1. The C# operator precedence table. Operators between rules have equal precedence.

You may not be familiar with all of these operators. The table is for reference, and throughout the course of your C# programming you will have an opportunity to learn about and use each operator.

Visual C# Reuses Many Operator Symbols

*There simply are not enough characters on a standard keyboard to represent all the operations that are possible in C#. As a result, C# must reuse some of the symbols, either by combining them with other symbols or by reusing them according to their **context**. The context is the meaning of an operator after taking into account the adjacent operators, variables, and values.*

For example, the shift operators are formed by combining two less than or two greater than symbols. The conditional OR and conditional AND operators are formed by combining two bitwise OR or two bitwise AND operators. The inequality operator is a combination of the NOT operator and the assignment operator.

When writing these operators, you should remember not to insert a space between the two symbols. If you do, the compiler will interpret the single operator symbols as two separate operations. This usually will cause the compiler to generate an error. Many times, however, it will not cause an error, but your program will not behave as expected. Finding such subtle errors can be difficult.

You may notice that the parentheses operator has three positions in the precedence table. The parentheses may be used to hold the argument list in a function call, to divide an expression and thus change the order of expression, or to cast one data type to another. The Visual C# compiler will determine how the parenthe-

ses will be used from the context of the program. This is not unlike a spoken or written language, in which the meaning of a word depends on the context in which it is used.

C# already uses most of the symbols on the keyboard that are not letters or numbers. Without reusing the symbols, C# would have to resort to a confusing array of operators using words instead of symbols.

Writing Arithmetic Expressions

Arithmetic expressions are those expressions that yield a numeric value. You can use the resulting number for any purpose. For example, it might just be a plain number that represents a temperature, a distance, or the elevation of a city.

To write an arithmetic expression, you must use at least one constant value or variable. If you use more than one value or variable, you must use an operator to tell the C# compiler what action you want to perform on the values or variables. You must *explicitly* write the operator.

In high school algebra, you probably learned that you can represent the multiplication of two variables by writing them adjacent to each other. For example, if you have variables *a* and *b*, you write the product as *ab*. The multiplication operator is *implicit*. That does not work in programming. The compiler would interpret *ab* as a variable other than *a* or *b*. In C#, and most programming languages, you must write *a * b* to indicate that you want the program to multiply *a* by *b*.

You can use arithmetic expressions in an assignment statement. This is a common use for assignment statements. In an assignment statement, an expression can use a variable that the program already has evaluated, as in the following example:

```
int pre_count = 8;
int count = 3 * pre_count + 2;
```

In this case, your program will evaluate the expression *3 * pre_count + 2* before assigning the result to the variable *count*.

You also can write an arithmetic expression as an argument to a function call. You will do this in the sample programs in Lesson 22, "Getting Started with C# Functions." Your program will not pass the expression to the function. Instead, your program will evaluate the expression fully as an *argument* to the function. After your program has evaluated the expression, it will pass a single value, the *parameter*, to the function.

The identifier on the left side of the assignment operator is the *lvalue* where "l" stands for the first letter of "left," and literally means the identifier on the left side of the assignment operator, the equals sign, or one of the other assignment operators. This identifier must be *modifiable*. It must be a variable in which you can store a value. For example, *5 = 2 + 3;* would cause the compiler to generate and display an error. The *5* is a constant, and is not a memory location where you can store the results of *2 + 3*. Instead, you would have to write something like the following:

```
int var;
var = 2 + 3;
```

The identifier *var* describes a memory location where your program can store the result of the expression.

The expression on the right side of the assignment operator is called the *rvalue*, but you will see that term used very infrequently.

Changing the Order of Expressions

You might have learned in elementary algebra that you can change the order in which you evaluate an expression by writing part of it inside a set of parentheses. The same is true in C#. Consider the following expression:

```
result = 4 + 7 * 6;
```

*Your program will perform the 7 * 6 multiplication first, then add 4 to the value and assign the value 46 to the variable **result**. However, there are times when you really want to do the addition first, then the multiplication. You could rewrite the expression as follows:*

```
result = (4 + 7) * 6;
```

*A C# program always evaluates the portion of an expression inside the parentheses first. In this case, your program will assign the value 66 to **result**.*

*In the **C2F.cpp** program, you converted degrees Celsius to degrees Fahrenheit. To work the formula the other way, you must use a set of parentheses to isolate the 32 degree difference between the freezing point in Celsius and the freezing point in Fahrenheit. The following program, **F2C.cpp**, uses parentheses to make sure the program performs the subtraction first:*

```
namespace nsTemps
{
    using System;
    class clsMain
     {
        static public void Main (string [] args)
        {
            if (args.Length == 0)
            {
                Console.WriteLine ("Please enter a " +
                            "temperature in Celsius");
                return;
            }
//  Convert the argument to an integer. A non-number
//  will throw a FormatException
            int Fahr;
            try
            {
                Fahr = Convert.ToInt32(args[0]);
            }
            catch (FormatException)
```

```
    {
        Console.WriteLine ("Please enter a " +
                        "numeric value");
        return;
    }
    int Celsius = 5 * (Fahr - 32) / 9;
    Console.WriteLine ("{0} degrees Fahrenheit " +
                        "= {1} degrees Celsius",
                        Fahr, Celsius);
    }
  }
}
```

This program evaluates **Fahr – 32** first because the expression is surrounded by parentheses. Without the parentheses, the equation would give a very different, and incorrect, result.

You can **nest** parentheses as well. Nesting parentheses involves writing one expression in parentheses inside another set of parentheses, as in the following statement:

```
result = 2 * (6 * (4 + 7) + 42);
```

Your program will evaluate the expression inside the innermost parentheses, 4 + 7, first. Then, your program will evaluate the expression within the outer set of parentheses using the result of the first expression, making it 11 * 6 + 42. Finally, your program evaluates the rest of the expression and assigns **216** to **result**.

Writing Boolean Expressions

Expressions do not necessarily have to return a number as the result. You often will need to know whether the result of an expression is 0 or not 0. At other times you might need to know the relationship between two expressions, whether the result of one expression is equal to, greater than, or less than the other expression.

To express these results, you use a mathematical system known as Boolean algebra which was invented by George Boole, an English mathematician. Boolean algebra often is used to describe the result of a group of switches or transistors in a circuit. In programming, you use Boolean algebra to combine the results of expressions, and return either a *true* or a *false* result. The C# *bool* data type holds this type of value.

Many of the C# operators return a Boolean result. For example, if you write *2 < 5*, you cannot represent the result as a numerical value. You are asking a question, "Is 2 less than 5?" Your program responds by returning *true*, or *1*, as the result of the expression. Of course, that result is obvious, but suppose you wanted to compare two variables that represent the results of two different operations. You would write *var1 < var2*, or "Is *var1* less than *var2*?" The result—*true* or *false*—will give your program the answer.

Using Boolean expressions, your program has the ability to make decisions. You will learn about decision-making statements in Lesson 20, "Making Decisions Within a C# Program." Earlier in this lesson, you saw an example of how to use a Boolean expression to determine whether the user had entered a temperature value on the command line when running the program:

```csharp
if (args.Length == 0)
{
    Console.WriteLine ("Please enter a " +
                       "temperature in Celsius");
    return;
}
```

Here the program is asking "Is the number of elements of the *args* array equal to 0." If the result is *true*, the program will execute the statement that follows, in this case printing a message for the user and exiting the program. If the answer is *false*, the program will ignore the statements.

In this lesson, you learned about expressions in C#, including how C# evaluates expressions and uses the results. An expression is a combination of operators, variables, and values that produce a result or cause an effect in C#. You learned that in most cases, C# will not let you discard the results of an expression. You also learned that C# assigns a precedence to its operators to assure that your program evaluates expressions in an orderly manner and consistently. Finally, you learned about Boolean expressions that return only *true* or *false* as a result. In Lesson 19, "C#, Like All Languages, Follows Rules of Grammar and Syntax," you will learn the grammatical rules of the C# language. Before you continue with Lesson 19, however, make sure you have learned the following key concepts:

- Your program uses operators, variables, and values in *expressions* as part of your program statements.

- A C# program evaluates an expression completely before using the result. It can use the result in another larger expression, or assign the result as the value of a variable. C# generally does not let you ignore the result of an expression.

- To make sure a program evaluates expressions properly, each operator has a precedence, as listed in the C# operator *precedence table*.

- You can modify the order of evaluation by writing parts of an expression in parentheses. C# programs evaluate expressions within parentheses first.

- Boolean statements use logic to evaluate an expression and return a *true* or *false* result. The C# *bool* data type holds a *true* or *false* value only.

int iCount = 0;
JumpToHere:
Console.WriteLine ("iCount = " + iCount);
++iCount;
if (iCount < 5)
    goto JumpToHere;

C#, LIKE ALL LANGUAGES, FOLLOWS RULES OF GRAMMAR AND SYNTAX

e very language—spoken, written, or computer—has rules that govern how you can combine its elements to produce proper expressions and statements. These rules form the language's grammar and syntax, and when you obey these rules you can create an infinite number of valid statements. C# inherits many grammatical rules from C++, but it has many of its own. You have learned how C# evaluates expressions, and in this lesson you will learn how C# applies its rules to the keywords, operators, and identifiers in your program. By the time you finish this lesson, you will understand the following key concepts:

- A language's *grammar* describes the overall structure and determines how you use punctuation marks and keywords.

- The *syntax* of a language deals with the structure of individual statements and determines how you use operators, numbers, and identifiers together.

- If your C# program does not obey the grammar and syntax rules of the C# language, the compiler will issue an error and will not compile your program.

- The syntax determines how your C# program will evaluate expressions within statements.

- The Visual Studio uses syntax to highlight the various elements in a C# program. You may customize the *syntax highlighting*.

C# Definitions and Conventions

There is a tendency to use the words "grammar" and "syntax" interchangeably. Syntax relates to how you assemble operators, variables, and identifiers to build a complete statement. Grammar is a broader term that describes the overall structure of the language and involves such things as how you define a namespace or class, or how you write a function. Technically, syntax is a *subset* of grammar, but it is a distinct entity in itself.

Certainly, any language has a finite number of words and punctuation marks, but it is a characteristic of any language that you may use these elements to build an infinite number of combinations. In programming, if that were not so, eventually all possible programs would have been written.

In addition, in any language there must be some agreement on basic concepts or no communication can take place. In a programming language, no programming could take place.

C# has some formal conventions that make up its syntax and grammar. You already may have seen some of these rules in C++:

❶ Whitespace separates identifiers such as constants and the names of variables.

❷ C# statements end with a semicolon—which is equivalent to writing a period at the end of a sentence.

❸ To insure orderly and consistent evaluation of expressions, C# assigns all operators a precedence.

❹ Identifiers written together must be joined by an operator. For example, to write "a times b," you must write $a * b$ and not simply ab or $a\ b$.

❺ C# does not let you discard the result of an expression, which would be akin to a sentence fragment. To store a value or the result of an expression, you must assign the value to a variable.

In a spoken or written language, the rules are open to interpretation by the listener or reader. You can make a mistake, but the listener or reader may be able to determine what you really mean. In C# *syntax* is more rigorous. The compiler will not try to determine what you really meant to write. If you make a mistake in writing a statement according to the syntax rules, the compiler will generate and display a syntax error and will not compile your program.

Understanding Operator Syntax

In C#, syntax is the set of rules that you use when you write program statements. The syntax determines how you use tokens and whitespace. You have learned that whitespace is any character that only moves the caret but does not display on your screen. You also learned that C# ignores whitespace except to separate tokens, so that you may use multiple whitespace characters between tokens.

Yet there are places where whitespace will confuse the compiler and cause it to generate a syntax error. For example, you have learned that C# contains a number of operators made from more than one symbol, such as the increment operator, ++. If you insert whitespace between these characters, the compiler will recognize them as separate tokens, two plus signs. When you write "+ +" (with a space between the symbols) in front of an identifier, the syntax is correct but the meaning is incorrect. The compiler recognizes two proper unary plus signs. When you write the symbols with a space between them after an identifier, the compiler almost certainly will generate a syntax error.

This is necessary, of course, because the C# syntax lets you apply a unary arithmetic operator to any expression that equates to a number, even if you use another unary operator in the expression. Try the following snippet in a test program:

```
int a = 42;
int b = + - 42;
Console.WriteLine ("b = " + b);
```

In evaluating the expression, your program first applies the unary minus sign to 42, then applies the unary plus sign to that result. The statement then assigns the result, *-42*, to *b*.

You should be aware of this quirk when writing program code. If you have a variable that does not appear to increment or decrement when you run your program, check to make sure you did not include a space accidentally.

The quirk appears only when you use the increment and decrement operators as prefix operators. When you use them as postfix operators, the compiler will generate and display syntax errors when you compile your code. The following lines will generate syntax errors:

```
count + +;
count+ +;
count - -;
count- -;
```

These statements cause syntax errors because, according the C# syntax, you must write an operand to the *right* of the unary plus or unary minus operators.

Reviewing Punctuation

Visual C# has very few punctuation marks, and you already have met most of them. "Punctuation" derives from the Latin word *punctus*, meaning point. Generally, the punctuation marks help the compiler to determine the intent of your programming.

The semicolon, for example, marks the end of a statement. By requiring this part of the language syntax, you may write long statements over several lines to improve the readability of your code. The semicolon thus becomes a point (*punctus*) that the compiler recognizes as the end of your statement.

Other punctuation marks signal the beginning and end of parts of your code. The open bracket marks the beginning of an expression that will equate to a number, which eventually will be used as a subscript for an array element. The close bracket marks the end of the expression. Table 19.1 summarizes the C# punctuation marks.

Punctuation Mark	Name	Purpose
[]	Brackets	Indicate array subscripts.
()	Parentheses	Group expressions, isolate conditions and indicate function calls.
{}	Braces	Indicate the beginning and end of a compound statement and mark the beginning and end of a class, structure or namespace definition.
,	Comma	Separates the parameters or arguments in a function definition or function call.
;	Semicolon	Terminates statements.
:	Colon	Indicates labeled statements used as the target for *goto* statements.

Table 19.1. C# punctuation marks and their uses.

You must use the first three punctuation marks in the table in pairs. Whenever you use an open bracket, open parenthesis or open brace, you must include the matching close mark.

Declaration Syntax

As you learned earlier, you cannot use a variable until you *declare* it. The declaration tells the compiler what data type to use for the variable's value and how much memory to set aside to store the value.

At some point in your code, you also must declare and define functions before you can use them. Unlike C++, the C# language makes no distinction between a function declaration and a definition. You must define the function by writing the body of code at the same point where you declare the function.

How you declare variables and functions is the language's *declaration syntax*. The statement in which you declare a variable is a *declaration statement*, which is simply a declaration of the data type and name of a variable followed by a semicolon:

```
int x;
```

You also may give the variable an initial value at the time you declare the variable by using an *assignment declaration*:

```
int x = 0;
```

The C# rules also let you declare multiple variables and give them initial values in a single statement. You must separate the variable names with a comma and the variables must be the same data type. They must be all type *int* or all type *long*, for example. You cannot declare an *int* and a *long* in this type of declaration:

```
int x = 42, y = 16, z;
```

All the variables in the preceding statement are type *int*. You should notice that the *z* variable does not have an initial value. You did not have to assign an initial value to all the variables in this type of assignment declaration.

C# uses a *definite assignment* rule to avoid program errors. The compiler will analyze your code to assure that you have assigned a variable a value before using it. In the case of a variable that you declare in a function, you must give a variable a value in an assignment statement or an assignment declaration. For variables that you declare as *fields* of a class, the compiler will assign a default value of 0 or *null* to the variable. If the compiler cannot find a default assignment or you do not initialize a variable, it will issue an error.

In addition, C# uses *proximity declaration*, which means that you may declare variables in a function at any place in the function *before* you use the variable. Proximity declaration helps to make your code more readable by letting you declare variables near the statements where you first use the variables.

Evaluating Expressions

An expression is a combination of one or more variables, values or operators. When your C# program evaluates an expression, you must use the result in another operation, such as an assignment operation or in a return statement.

The C# compiler will not let you ignore the results of an expression and will issue an error if you do not use the value in an operation.

This is a major departure from C and C++, which let you ignore expressions. Early C and C++ code often was used with other modules written in Assembly. Programmers sometimes used the results of an expression to set the various *flags* in the computer's processor chip before calling a function written in assembly. When used this way, the actual result of an expression was not as important as its effect, so C and C++ let you discard the results. C#, however, is not designed to interface with assembly code, so a discarded expression is simply wasted code.

The *data type* of an expression will be the highest precision of any value used in the operation. In other words, if you mix a *float* data type with an *int* type in the same expression, the result of the expression will be a *float* type value. C++ automatically *promotes* values to the next higher precision in an expression. The expression *6.3 / 2* returns a *float* value although one of the two values, the *2*, is an integer:

Your program will evaluate subexpressions before promoting the value, and this can lead to some unexpected results. If you evaluate the following snippet on a calculator, you will find the result, the value assigned to *y*, is 63.45:

```
int x = 3;
double y = 42.3 * (x / 2);
Console.WriteLine ("x = {0,0:F2}", x);
```

However, when you add these lines to a test program and print or examine the result, you find that the value of *y* actually is 42.3. C# evaluates an expression within parentheses first. Both the *x* and the *2* within the parentheses are integers, so your program performs integer division, and the result is 1.

You can force an expression such as this to a *double* by writing one of the operands as a *double* value, either by casting the *x* to a *double* or writing the divisor as *2.0*, as shown in the following snippet.

```
int x = 3;
double y = 42.3 * (x / 2.0);
Console.WriteLine ("x = {0,0:F2}", x);
```

When your program examines this expression, it first will cast the variable *x* to a *double* before evaluating the expression, giving the result of 63.45.

Writing Statements

Statements control the flow of your C# program's execution. So far, you have learned about statements that evaluate expressions or assign values to a variable. Visual C#, however, has several statement types that let you perform *loops* or transfer control to other statements.

Table 19.2 lists the various statement types in Visual C#. You may not be familiar with most of these yet, but during the course of this book you will learn about all the statement types and use them in programs.

Statement Type	Purpose
break	To break program execution out of a loop statement.
continue	To restart a loop before all the statements in the loop have executed.
do ... while	To enclose a compound statement inside a loop that will execute at least once.
for	To repeat a statement in a *loop* a specified number of times.
foreach	Similar to a *for* statement but steps through the elements of an array.
goto	To cause the program to jump to a labeled statement.
if	To provide condition execution of another statement.
label	To provide a target for a *goto* statement. The label name must be followed by a colon, as in *Name:*
null	Null statement. Contains only a semicolon but no expression.
return	To transfer control from a function to another function that called it.
switch	To provide several optional execution points.
try	To provide a guarded block to execute statements that may throw exceptions.
while	To repeat a statement in a *loop* until a condition becomes *false*.

Table 19.2. C# Statement types

In addition to these special statement types, a statement may be *expression* or *compound*. Expression statements sometimes are called *simple* statements, and contain a single terminating semicolon. It may be a *null* statement that contains only a semicolon and no expression. A compound statement contains one or more expression statements between open and close braces ("{" and "}"). Often you will see the word "block" used to refer to compound statements. Compound statements have additional properties that control the *scope* or visibility of variables.

The *try* and *catch* statements are examples of compound statements. You must write the guarded code in a *try* block as a compound statement. Then, to catch an exception, you must write the alternate code inside a *catch* block. You used this construction in the *C2F.cs* and *F2C.cs* programs in Lesson 18 to protect the code against an invalid argument:

```
try
{
    Fahr = Convert.ToInt32(args[0]);
}
catch (FormatException)
{
    Console.WriteLine ("Please enter a " +
                        "numeric value");
    return;
}
```

If the program user had entered something other than a number—say, the word "foo"—as a temperature, the *ToInt32()* function would throw a format exception. If this happens, the program transfers control to the *catch* block. As punctuation marks, the braces mark the beginning and end of these blocks of code.

You will learn more about compound statements in the next two lessons. In addition, you will learn how compound statements affect the *scope* of variables in Lesson 25, "Understanding Function and Variable Scope."

Using the null, goto and label Statements

Three entries in Table 19.2 have limited use and will not be covered in later lessons. These are the **null**, **goto** and **label** statements.

The **goto** statement **unconditionally** transfers control to another statement within the same function. You use the **goto** statement with a **label**, which is a unique name similar to the name of a variable except that it ends with a colon:

```
int iCount = 0;
JumpToHere:
    Console.WriteLine ("iCount = " + iCount);
    ++iCount;
    if (iCount < 5)
        goto JumpToHere;
```

The **goto** statement has had a lot of bad press among programmers almost from the beginning of the computer age. Many programmers claim that it breaks the program flow. Usually, it is possible to find other programming constructs to replace the **goto**, but often you will find it handy as a debugging device.

The object of a **goto** must be a label statement, and the label must be in the same function as the **goto** statement. You cannot transfer control to a statement in another function. In addition, the label is **local** to a function, which means that other functions do not even know that it exists. The label only marks a statement, and you must follow a label with a valid C# statement.

The null statement is nothing but a semicolon written by itself. It does not calculate anything, it does not yield any result and it does not assign a value to anything. From its name, you might think that the null statement does not do anything, and you would be correct. That is not to say that a null statement is useless, however.

If you accidentally type two semicolons at the end of a statement, you actually are creating a null statement:

```
int x = 42;;
```

The first semicolon ends the statement and the second is a null statement. The null statement does nothing, but it is legal and the compiler will not issue an error.

Most commonly, other than accidentally creating a null statement, you use a semicolon by itself to mark a label statement where you have no code to execute:

```
void TestFunc()
{
    int iCount = 0;
    while (true)
    {
        Console.WriteLine ("iCount = " + iCount);
         ++iCount;
        if (iCount > 5)
            goto DoneWithLoop;
    }
DoneWithLoop: ;
}
```

Without the **goto** statement, this code would loop indefinitely. However, when **iCount** reaches a value of 6, the **goto** statement transfers control to the label statement. At the label, you have no code to execute, but the C# grammar requires that you provide a statement. The null statement satisfies this condition.

Using Syntax Highlighting in Visual Studio

The Visual Studio text editor uses the C# language syntax to display parts of your program in different colors and fonts. The Visual Studio's use of color should help you to identify the various statements in your program and to locate specific elements. This *syntax highlighting* is a feature of the Visual

Studio and the Visual Studio editor does not save the colors and different fonts as a part of your source-code file.

To demonstrate syntax highlighting, open a Visual Studio project. The *F2C* project from Lesson 7 will provide a good example. When you open the project, display the *Class1.cs* file by double-clicking your mouse on the entry in the Solution Explorer.

In the *Class1.cs* file, notice that the Visual Studio editor displays the C# keywords in blue. This is an example of syntax highlighting. It helps you to identify the keywords, and helps you to identify errors. For example, if you type **clsss** instead of **class,** the word does not change to blue when you type it, and you know immediately that you made a typing error.

You can customize the syntax highlighting in Visual Studio. Select the Tools menu, then select the Options item. This will open the Options dialog box. Click on the Environment item in the box on the left side of the dialog box, then click your mouse on Fonts and Colors. The dialog box should be similar to that in Figure 19.1.

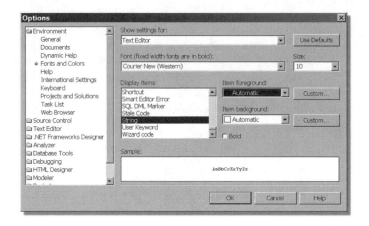

Figure 19.1 The Visual Studio Options dialog box showing controls to set syntax highlighting.

A common error in dealing with strings is to forget a close quote at the end of the string. By default, the Visual Studio displays strings in the same color as other text in the program file. Change the syntax highlight for strings to display them in red by following the following steps:

❶ With the Options dialog box displayed, select Environment, then Fonts and Colors. Make sure the Show Setting For box displays "Text Editor."

❷ Use the scroll bar on the Display Items box to find the entry for String. Click your mouse on this item to select it.

3 Click your mouse on the arrow next to the Item Foreground box to display a list of preset colors. Using the scroll bar on this list, select Red. The Sample box will show an example of text in the color you select.

4 Click your mouse on the OK button to make the change.

In the *Main()* function, notice that the string portion of the calls to the *Console.WriteLine()* function now display in red. Type the following line, intentionally leaving out the ending quote mark:

```
Console.WriteLine ("The answer is {0}, 42);
```

The number 42 displays in red as well, immediately notifying you that something is not right. Add the closing quote mark before the comma and the number will display normally.

Return to the Options dialog box and change the color for the Number entry to another color, such as magenta or green. When you click OK, the 42 will appear in the new color. Also, try changing the Operator and Identifier entries to different colors.

By setting different colors for the various parts of a program file, you can quickly move about the source code to identify errors and debug code.

WHAT YOU MUST KNOW

In this lesson, you learned about C# grammar and syntax and how your C# program evaluates expressions. Grammar describes the overall structure of a language. The syntax describes how you use individual values, identifiers and operators in a single statement. You must obey the grammatical and syntax rules of C# or the compiler will issue an error and refuse to compile your program. You also learned how to use syntax highlighting in the Visual Studio to make elements in your program stand out. In Lesson 20, "Making Decisions Within a C# Program," you will learn how to use the *if* statement to give your programs the ability to make decisions. Before you continue with Lesson 20, however, make sure you have learned the following key concepts:

- C# uses grammatical rules to determine the overall structure of a program, and syntax rules to guide you when you write individual statements.

- The grammar rules determine how you declare and use classes, structures and namespaces in your program.

- The syntax rules determine how you declare variables, use operators, and write expressions.

- Using *proximity declaration*, you may declare variables in a function near the point where you first use the variables.

- C# contains a number of statements that let your program loop, make decisions and alter the program flow.

- Using the Options dialog box, you may customize the Visual Studio to display program elements in different colors and fonts.

MAKING DECISIONS WITHIN A C# PROGRAM

a computer is a machine. Certainly, it uses electronics to determine its operation, but it is a machine nevertheless. What distinguishes it from other machines is the fact that it can make *decisions*. The engine in your car, for example, cannot determine whether the pistons should go up or down; that is determined solely by the position and orientation of the crankshaft. The engine cannot determine whether a valve should open or close; the camshaft determines the valve operation (although, it should be noted, many engines use computers to make these decisions). A program running on a computer, however, can examine the program environment (or, with the help of external sensors, the physical environment) and decide where the program flow should go. In this lesson, you will examine the decision-making constructs available to you in C#. By the time you finish this lesson, you will understand the following key concepts:

- Using *conditional processing*, you can make your program perform or ignore certain statements depending upon the result of a conditional test.
- The *if* statement is the primary statement for making a decision in C#. The *if* statement requires a *conditional expression* followed by a *conditional statement*. A program will execute the conditional statement only if the condition expression is *true*.
- The *else* statement lets your program execute an alternate statement or block of code if a conditional test fails.
- You may combine multiple conditional tests using the *else if* construct.

- You may nest *if* statements by providing another *if* statement in the conditional statement.
- Within C#, the switch statement, which lets you perform specific processing based on the result of an expression, behaves differently than the switch statements you may have used in C and C++.

Understanding Decision-Making Constructs

A program is a list of instructions that your computer executes one after another. The compiler turns your source code into the numbers that the computer understands as instructions and data. Some of these computer instructions test certain parts of the program environment—such as the value of a variable—and tell you the results of that test. By using the results of these tests, your program can execute or ignore certain blocks of code, or even execute an alternate block of code.

Programs that make decisions perform *conditional processing*. To make a decision, your program performs a test on values in your program. Using the results of the tests, you can avoid executing code that causes your program to crash, or you can instruct the user to enter some needed information.

For example, your store's policy may require a clerk to ask a customer for identification for checks over $20. At the end of the transaction, your program could test the total value of the sale and display a reminder to the clerk.

C# provides a number of operators that permit your program to test the relationship between two values, as summarized in Table 20.1. The first two operators in the table are the *equality* operators and the rest are the *relational* operators. The equality operators and several of the relational operators use more than one symbol, and you must remember to write the symbols with no spaces between them. You can write spaces before or after the operators, but not between. For example, the "is equal to" operator is two equals signs written together as in *val1* == *val2*. If you write this as *val1* = = *val2*, the compiler will interpret that as two equals signs and issue a syntax error.

Operator	Test	Example
==	If two values are equal	(value == 50)
!=	If two values are not equal	(value != 50)
<	If the first value is less than the second	(value < 50)
>	If the first value is greater than the second	(value > 50)
<=	If the first value is less than or equal to the second	(value <= 50)
>=	If the first value is greater than or equal to the second	(value >= 50)

Table 20.1. The Visual C# equality and relational operators

Expressions written using the equality and relational operators return only *Boolean* values; that is, the result is either *true* or *false*. Most of the conditional expressions in C# require a Boolean value, which is a major difference between C# and C++. For example, in C++ you can use a variable in a conditional expression, and the expression will return *true* if the value of the variable is not zero or *false* if the value is zero as in the following:

```
int x = 10;
if (x)
    . . .
```

That will not work in C# unless you have declared *x* as a Boolean variable. You must specifically test the variable to see if its value is zero:

```
if (x == 0)
    . . .
```

If you make this sort of error in C#, the compiler will tell you that it cannot "Explicitly convert type 'int' to 'bool'."

Using the *if* Statement

The primary decision-making statement in C# is the *if* statement. The *if* statement requires a conditional expression following the keyword and a conditional

statement that will execute when the expression is *true*. The conditional statement may be a simple statement, or it may be a compound statement written between a set of braces.

In its simplest form, the *if* statement determines whether an expression is *true* or *false*, as shown in the following snippet:

```
if (conditional expression) conditional statement;
```

The *conditional statement* is the *object* of the *if* statement. When the result of the conditional expression is true, the conditional statement will execute. Otherwise, your program will ignore the conditional statement. Usually, programmers write the conditional expression on one line, then the conditional statement on the next line with a four-space indent:

```
if (conditional expression)
    conditional statement;
```

Notice that the first line *does not* end with a semicolon. It is important to remember that the two lines together form a complete statement. The extra line helps keep you from overlooking the conditional statement, and the indent indicates that the second line depends on the previous line.

In the following program, *First_if.cs*, the program tests whether the value of *iScore* is *greater than or equal* to 90. If the value is 90 or higher, your program will execute the *Console.WriteLine()* statement following the test:

```
namespace nsFirst
{
    using System;
    class clsMain
    {
        static public void Main ()
        {
            int iScore = 95;
```

```
        if (iScore >= 90)
            Console.WriteLine ("Congratulations. " +
                                "You got an A!");
        }
    }
}
```

Compile and test-run the program, and you should see the congratulatory message appear on the screen. Try changing the value of *iScore* to something less than 90. Recompile and run the program. The program should run and exit without printing anything. The program ignores the conditional statement.

Very often, you must execute more than one statement when the result of a conditional expression is *true*. In this case, you may use a compound statement as the object of the *if* statement. You must start the compound statement with an open brace and end it with a closing brace as in the following snippet:

```
int iScore = 95;
if (iScore >= 90)
{
    Console.WriteLine ("Congratulations. You got an A!");
    Console.WriteLine ("Your score is {0}", iScore);
}
```

When written this way, your program will execute *all* the statements between the braces if the expression is *true*, or it will ignore all the statements. There is no intermediate ground.

You may reverse the *sense* of the conditional expression using the *not* operator, an exclamation point. In Lesson 18, "Writing Expressions in C#," you learned that the *not* operator has a higher precedence than the equality and relational operators, so you must write the conditional expression inside a separate set of parentheses and then apply the *not* operator:

```
int iScore = 95;
if (!(iScore >= 90))
```

```
{
    Console.WriteLine ("Sorry. You did not get an A.");
    Console.WriteLine ("Your score is {0}", iScore);
}
```

If you write the expression without the inner set of parentheses, the compiler will associate the *not* operator with the *iScore* variable and issue an error telling you that you cannot use the *not* operator with an *int* variable.

Using the *else* Statement

When your program performs an *if* statement, it will execute the object statement or block of statements. You can reverse the sense of the test using the *not* operator, and thus perform the object statement or block when the expression is *false*.

Sometimes, however, you must do both. You must execute statements when the test is *true*, or another set when the test is *false*. You could write two *if* statements, testing for both possibilities. However, C# provides a method of writing an alternate set of statements by using the *else* keyword.

By using the *if-else* sequence, your program will execute the statement after the *if* when the conditional expression is *true*, and the statement after the *else* when the conditional expression is false. The following short program, *IfElse.cs*, takes a number score from the command line, then prints one message if the grade is 90 or greater, or another message if the grade is less than 90:

```
namespace nsIfElse
{
    using System;
    class clsMain
    {
        static public void Main (string [] args)
        {
            if (args.Length == 0)
            {
                Console.WriteLine ("Please enter a number");
                return;
            }
```

```
            int iScore = 95;
            try
            {
                iScore = Convert.ToInt32(args[0]);
            }
            catch (FormatException)
            {
                Console.WriteLine ("Please enter a number");
                return;
            }
            if (iScore >= 90)
                Console.WriteLine ("Congratulations. " +
                                   "You got an A!");
            else
                Console.WriteLine ("You need to try " +
                                   "harder next time");
        }
    }
}
```

In *IfElse.cs*, you use the *if* statement first to check whether the user actually entered an argument on the command line. If *args.Length* is 0, you write a reminder message to the screen, then return from *Main()*, which ends the program. Then, using *try . . . catch* blocks, you make sure the user typed a number value. Finally, if the score is 90 or greater, you write a congratulatory statement, or, if the score is less than 90, you write a dunning message.

As with the *if* statement alone, you may use a compound statement after the *else* to execute more than one statement when the expression is *false*:

```
if (iScore >= 90)
{
    Console.WriteLine ("Congratulations. You got an A!");
    Console.WriteLine ("Your score was {0}", iScore);
}
else
{
    Console.WriteLine ("You did not get an A");
    Console.WriteLine ("You need to try harder next time");
}
```

Mixing Compound Statements with if-else

You may build compound statements by grouping simple statements inside a set of braces. If you must execute more than one statement when your test is **true**, you may use a compound statement after the **if** statement. Similarly, if your program must execute more than one statement following the **else** keyword, you may use a compound statement as well.

When you use the **if-else** construct, you do not have to use compound statements after both the **if** and the **else**. You may use a simple statement after one or the other, and a compound statement after the other as shown in the following sample code:

```
if (iScore >= 90)
    Console.WriteLine ("Congratulations. You got an A!");
else
{
    Console.WriteLine ("You did not get an A");
    Console.WriteLine ("You missed {0} points",
                        100 — iScore);
}
```

Building Multiple if-else Constructs

The *if . . . else* construct lets you provide one block of code to execute when a test is *true*, or a second block of code that will execute if the test is *false*. Not everything in life, however, can be expressed simply as true or false. Even testing a single variable, for example, presents you with three possibilities. The variable may be less than the test value, it may be equal to the test value or it may be greater than the test value. In the case of a grade, you may want to check for multiple values to give the score a letter grade.

You may test for multiple possibilities by "stacking" *else* statements and combining them with an *if* statement. The combined *else if* statement lets you provide another expression to test a value. The following program, *Grades.cs*, uses multiple *else if* statements:

```csharp
namespace nsIfElse
{
    using System;
    class clsMain
    {
        static public void Main (string [] args)
        {
            if (args.Length == 0)
            {
                Console.WriteLine ("Please enter a number");
                return;
            }
            int iScore = 95;
            try
            {
                iScore = Convert.ToInt32(args[0]);
            }
            catch (FormatException)
            {
                Console.WriteLine ("Please enter a number");
                return;
            }
            if (iScore > 100)
                Console.WriteLine ("That score is " +
                                        "impossible");
            else if (iScore >= 90)
                Console.WriteLine ("Congratulations. " +
                                        "You got an A!");
            else if (iScore >= 80)
                Console.WriteLine ("You got a B");
            else if (iScore >= 70)
                Console.WriteLine ("You got a C");
            else if (iScore >= 60)
                Console.WriteLine ("You got a D");
            else
                Console.WriteLine ("You failed the test");
        }
    }
}
```

When you have multiple tests using the *else if* statement, your program will test the conditions until it finds one that is *true*. If none of the expressions is true, then the statement after the *else* statement will execute. You do not need to include an *else* statement, but if you do, it must be the last statement in the "stack." The program will execute one and only one of the statements.

Using the *switch* Statement as an Alternative to Multiple *if-else* Statements

In the last topic, you learned how to combine the else and *if* statements to provide a new test expression. Then you learned how to "stack" *else if* combinations to test for multiple possibilities. When one of the tests returned *true,* the statement or compound statement for that test executed. Your program ignored all other statements.

The *Grades.cs* program tested for ranges of values. Sometimes you will be able to test for specific values rather than ranges of values. In this case, you may use the C# *switch* statement to simplify your code. To use the *switch* statement, you must provide a condition that evaluates to an *integral* value, which is a numerical value that is a whole number (it does not have any digits after the decimal point). Examples of integral data types are the *char*, *short*, *int* and *long*. The following program, *Switch.cs*, uses a *switch* statement to test for a letter grade:

```
namespace nsSwitch
{
    using System;
    class clsMain
    {
        static public void Main (string [] args)
        {
            if (args.Length == 0)
            {
                Console.WriteLine ("Please enter a grade");
                return;
            }
            char chScore = (char) (args[0][0] & 0xdf);
            switch (chScore)
            {
                case 'A':
                    Console.WriteLine ("Congratulations "+
```

```
                                 "on your A!");
            break;
        case 'B':
            Console.WriteLine ("B is OK, but you " +
                                 "can get an A");
            break;
        case 'C':
            Console.WriteLine ("C is an average " +
                                 "grade");
            break;
        case 'D':
            Console.WriteLine ("D is barely " +
                                 "passing");
            break;
        case 'F':
            Console.WriteLine ("You failed! " +
                                 "Study harder!");
            break;
        default:
            Console.WriteLine ("Your grade is " +
                                 "unknown");
            break;
        }
    }
}
}
```

When you use the *switch* statement, there are some important rules to remember:

❶ The test expression *must* be a value, variable or expression that results in an integer value such as *bool*, *int*, *long* or *short int*. You cannot use a *float* or *double* value for the test value. You also can use a string as the test expression.

❷ The *case* keyword marks the alternate blocks of code your program may execute. Your program will execute a block if the test expression equals the value after the *case* keyword. The *case* value must be a constant. You cannot use a variable or an expression containing variables after *case*.

❸ The statement following the *switch* statement *must* be a compound statement. Even if you provide only one *case* statement, you must include it as part of a compound statement. This means you must use the open brace to mark the beginning of the *switch* block and the close brace to mark the end of the *switch* block.

④ The *break* statement ends the compound statement after the *switch* statement. When your program encounters the *break* statement, your program will transfer control to the first statement after the closing brace. If you do not include the *break* in a block of code, you must provide a *goto* statement to jump to another *case* block.

⑤ Your program will execute the statements in the *default* block if the test value does not match any of the values in the *case* statements. This lets you perform statements for an unexpected value.

If you have programmed in C++, you should notice the difference in item 4. In C++, if you do not provide a *break* statement, the program will "fall through" to the next *case*. In C#, this will cause the compiler to generate an error. To jump to the next *case*, or even any other *case*, you must provide a *goto* statement, as in the following snippet:

```
case (x)
{
    case 1:
        // Statements
       goto case 3;
    case 2:
        // Statements
        break;
    case 3:
        // Statements
        break;
}
```

Be careful how you write the test expression. If you use one of the equality or relational operators, the result will be *Boolean* and can only be *0* or *1*. For example, if you write the test expression as *(chScore == 'A')* your *case* statement may be only *true* or *false*. You can see this in *Switch.cs* if you change the expression. The compiler will issue a series of error messages similar to the following for all the case statements other than *default*:

```
Switch.cs(25,22): error CS0029: Cannot implicitly
convert type 'char' to 'bool'
```

The *case 'A'* actually is a value of 65, which is not *true* or *false*. The compiler is looking for *case* statements containing Boolean values, and the only case that matches is *default*. Thus, the compiler issues the errors and will not compile your program until you correct the expression or the *case* statements.

Evaluating More Than One Expression in an if Statement

The ability to make decisions gives your program the ability to select alternate blocks of code depending upon the results of a conditional expression. Those decisions so far have been based on a single expression. Your decision may not always be so easy, however. Sometimes you must evaluate two or more expressions to get the result you need to make a decision.

Visual C# provides two *logical* operators that allow you to combine two Boolean, or logical, values. The logical operators require two operands, and both operands must be of type *bool*. Table 20.2 shows the C# logical operators.

Operator	Meaning	Example
&&	AND	*condition1 && condition2*
||	OR	*condition1 || condition2*

Table 20.2. Visual C# logical operators for combining Boolean values.

In the *Grades.cs* program earlier in this lesson, you tested the command line value to make sure it was not greater than 100. Suppose the command line value is a negative number. That would not be a valid test score either. You could provide two separate tests, one for greater than 100 and another for less than 0, but you would have to write separate statements to print to the console. Instead, you may combine the tests into a single conditional using the logical AND operator, as in the following snippet:

```
if ((iScore > 100) || (iScore < 0))
{
    Console.WriteLine ("That score is " +
                        "impossible");
    return;
}
```

By combining the conditional expressions, your program now rejects a negative score as the command line argument.

Notice that the two conditional expressions are written within parentheses. In Lesson 18, you learned that the equality and relational operators have a higher precedence than the logical operators, so the inner sets of parentheses are not needed. However, by grouping them within parentheses, the code is easier to read. Compare the previous snippet with the following:

```
if (iScore > 100 || iScore < 0)
{
    Console.WriteLine ("That score is " +
                       "impossible");
    return;
}
```

You have to stop and mentally associate the operands with the operators when you read the code. The extra parentheses do not add overhead to your program, but they do make the code easier to read.

Solving Logical Operations Using a Truth Table

*The logical operators are well known operations borrowed from the study of symbolic logic. Logic deals primarily with the manipulation of language, whether those languages are mathematical, spoken or computer programming. Because the operands of expressions using && and || must be logical **true** or **false** values, you may apply **truth tables** from logic to the operands to evaluate the expressions.*

Table 20.3 shows the truth tables for the logical AND and OR operators.

Value1	Value2	Value1 && Value2	Value1 || Value2
true	true	true	true
true	false	false	true
false	true	false	true
false	false	false	false

Table 20.3. The truth table for C# logical operators.

As you can see from the truth table, an expression combining operands using the && operator is **true** if and only if both of the operands are **true**. However, an expression combining operands using the || operator is true if either or both of the operands is true.

Nesting Decision-Making Statements (Placing One Statement Within Another)

The *if* statement requires another statement as its object. There are no restrictions on what type of statement you may use as the object. If it is a valid C# statement, it may be used as the object of an *if* statement.

Of course, this implies that you may use another *if* statement as the object of an *if* statement, and this is the case. The following program, *Nested.cs*, shows how one *if* statement may be nested in another:

```csharp
namespace nsNested
{
    using System;
    class clsMain
    {
        static public void Main ()
        {
            int iScore = 85;
            if (iScore < 90)
            {
                if (iScore >= 80)
                {
                    Console.WriteLine ("Work harder. " +
                                       "You got an B.");
                    Console.WriteLine ("Your score was {0}",
                                       iScore);
                }
            }
        }
    }
}
```

When this code executes, it first will test whether *iScore* is less than 90. If the score is 90 or above, the student got an "A" and the compound statement following does not apply. If the score is less than 90, the program will then test whether the score is 80 or greater. Because the score is 85, the *Console.WriteLine()* statements will execute.

WHAT YOU MUST KNOW

In this lesson, you learned how your program can evaluate the truth value of an expression. You then can use this truth value, *true* or *false*, in an *if* statement to make a decision. Decision making gives your program the power to execute or ignore portions of code, or to provide alternate statements to perform based on the results of a logical expression. You learned how to combine condition expressions, and how to nest *if* statements. In Lesson 21, "Repeating Statements within a C# Program," you will learn how to use decision-making constructs to make your program repeat statements, or to perform a *loop*. Before you continue with Lesson 21, however, make sure you have learned the following key concepts:

- You use the *if* statement to make a decision based on the Boolean value of a test expression.

- The test expression in C# *must* evaluate to a value of type *bool*.

- When the result of the test expression is *true*, your program will execute the statement immediately following. The statement may be a simple statement or it may be a compound statement.

- If the result of the test is *false*, your program ignores the statement immediately following. However, you may provide an alternate statement using the *else* keyword.

- The *else if* combination allows you to specify an *alternate* condition when the *if* statement is *false*.

- You use the *switch* statement to select one of several blocks based upon the *value* of a variable or constant. Under C#, the switch statement no longer allows fall-through operations.

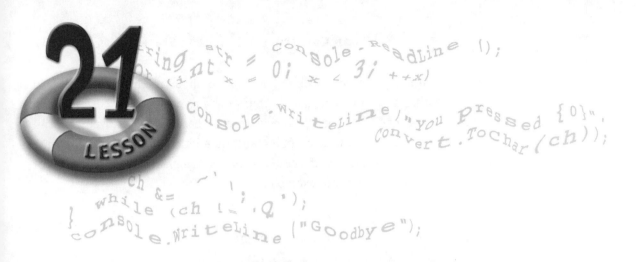

REPEATING STATEMENTS WITHIN A C# PROGRAM

ecision-making statements using the *if* keyword give your programming efforts a powerful construct. Now your programs can branch and execute certain portions of code depending upon the current value of variables. Another powerful programming construct is the *loop*, which lets your program execute a statement or block of statements repeatedly. Looping involves decision making that is not as obvious as the *if* statement. Somewhere in your repeating code, your program must decide when the loop has served its purpose, and force the loop to exit. Visual C# contains several loop statements. In this lesson, you will examine the various C# loop statements and learn how to use them. By the time you finish this lesson, you will understand the following key concepts:

- Looping involves making your program repeat a statement based on the value of a control expression.
- In most forms of loops, you declare and initialize a control variable. You use the control variable in the control expression.
- To make your program execute a loop a given number of times, you use the *for* loop.
- The *while* loop will execute indefinitely until the control expression evaluates to *false*. A variant of the *while* loop is the *do-while* loop, which assures your loop statements will execute at least once.
- C# contains a special form of the *for* loop called the *foreach* loop. You use the *foreach* loop with arrays.
- As with decision-making statements, you may nest loop statements.

Using C# Loop Constructs to Repeat Statements

The ability to make decisions, which you learned about in Lesson 20, "Making Decisions within a C# Program," means your programs can test the current value of variables and perform alternate blocks of code based on those values. Still, the *if* statement only lets your program perform a block of code once and only once. Many times you must perform a block of code repeatedly, to initialize or read the values of an array, for example.

You perform repeated execution of a statement or block of statements using *loops*. Looping is an important concept in programming.

Looping involves transferring control to a statement that the program already has executed. You may want to do this when you want to repeat a group of statements until a certain condition occurs. For example, your program might open a data file, read a line, and process the line. When your program has finished processing the line, it might return control to the statement that reads a line. This would continue until your program has read the entire file. At that point, it would transfer control to a statement outside the loop.

In Visual C# there are four different loop control statements, as summarized in Table 21.1.

Loop Statement	Purpose
for	Executes a statement or block of statements a specific number of times. It is similar to the FOR ... NEXT loop in BASIC.
foreach	Executes a block of statements for every element in the array.
while	Executes a statement or block of statements indefinitely as long as a condition is *true*.
do while	Executes a statement or block of statements *at least once*, then indefinitely as long as a condition is *true*.

Table 21.1. The Visual C# loop statements.

Each of these loop statements involves writing a *control expression*, which your program uses to decide whether to repeat a statement or block of statements. Each time your program executes the statement or block of statements in a loop is an *iteration* of the loop. You will find loops are handy devices to improve the efficiency of your code, and it is important that you fully understand how to use these statements.

Looping with *goto* and Label Statements

The most basic method of building a loop in C# is to use the *goto* statement with a *label* statement. It also is the most instructive. A label, you have learned, is a unique name within a function that serves as the target for a *goto* statement. All the C# loop constructs may be written in a form that uses a *goto*. Understanding this basic loop form will give you some insight as to how the various loop statements operate.

The following short program, *GotoLoop.cs*, builds a loop using a label with a *goto* statement:

```
namespace nsLoop
{
    using System;
    class clsMain
    {
        static public void Main ()
        {
            int iCount = 0;

        JumpToHere:
            Console.WriteLine ("iCount = {0}", iCount);
            ++iCount;
            if (iCount < 5)
                goto JumpToHere;
        }
    }
}
```

The statement following the label *JumpToHere* is the first statement in the loop *body*. The label marks the beginning the loop, and all of the statements from that point to the *if* statement will execute each time the program repeats the group of statements.

The *if* statement is the decision point. That is where your program decides whether to return to the label and repeat the code. The *iCount < 5* is the *control* *statement*. As long as the control statement returns *true*, the code will repeat. The variable *iCount* is the *control variable*. Its value determines the result of the test in the control statement. This construction is very similar to the way the C# *for* loop operates.

When you compile and run *GotoLoop.cs*, you should see the following output:

```
iCount = 0
iCount = 1
iCount = 2
iCount = 3
iCount = 4
```

The control statement specifies that *iCount* must be less than 5 to repeat the loop, so when *ICount* reaches 5, the program ignores the *goto* statement and the loop ends.

You could rewrite the program using the label statement at the top and a compound statement as the object of the *if* statement, as in the following program, *Goto2.cs*:

```
namespace nsLoop
{
    using System;
    class clsMain
    {
        static public void Main ()
        {
            int iCount = 0;

        TopOfLoop:
            if (iCount < 5)
            {
                Console.WriteLine ("iCount = {0}", iCount);
                ++iCount;
                goto TopOfLoop;
            }
        }
    }
}
```

The compound statement executes only if *iCount* < 5. In the compound statement, you change the value of the control variable, *iCount*, and transfer control to the top of the loop without further testing. This is similar to the C# *while* loop.

Using the *for* Loop to Perform Statements a Specific Number of Times

You often will find it convenient or necessary to perform a block of statements a specific number of times. You may need to execute one or more statements for each element in an array, or you may need to print five copies of a file. Of course, you could build a loop consisting of a *goto* and a label as in the previous section. However, the Visual C# *for* statement makes it very easy to repeat a block of statements a specific number of times.

To use the *for* statement (usually called a *for* loop), you usually will need a *control variable* along with a *control expression*. The loop will execute until the control variable's relationship to another value in the control expression becomes false. Normally, you will use the control variable to keep track of the number of times the loop executes.

The *for* loop in the following snippet will execute 10 times:

```
for (iCount = 1; iCount <= 10; ++iCount)
    statement;
```

The *for* loop consists of four parts. The first three are the loop control statements, which you write inside a set of parentheses immediately after the *for* keyword. The first statement, *iCount = 1* in the previous snippet, is the initializing statement. The loop will execute this statement only once, just before the loop begins. You may use this statement to initialize the loop control variable. You end this statement with a semicolon.

The second part is the *control expression*, which is *iCount <= 10* in the snippet. Your program will evaluate this expression before each iteration of the loop. If the expression is true, the loop will execute. If the expression is false, the loop will not execute and the program control will transfer to the first statement after the loop.

Thirdly is the *increment*. Despite its name, you may write virtually any arithmetic expression in this statement. Your program executes this statement after each iteration of the loop. Normally, you would modify the control variable in this statement.

After the increment, you write a close parenthesis. After that is the body of the loop. The body may be a simple statement, or it may be a compound statement enclosed in a set of braces:

```
for (iCount = 1; iCount <= 10; ++iCount)
{
    statement1;
    statement2;
}
```

It is good practice always to write the loop statement as a compound statement, even if you have only one statement to execute. This helps you to identify the loop and isolate the statements when debugging a program. Later, if you add another statement to the loop, you will not have to worry about forgetting to add the braces.

The following short program, *FirstFor.cs*, uses a loop to print the value of *iCount* each time the loop executes:

```
namespace nsForLoop
{
    class clsMain
    {
        static public void Main ()
        {
            int iCount;
            for (iCount = 0; iCount <= 5; ++iCount)
            {
                Console.Write ("iCount = {0}", iCount);
            }
        }
    }
}
```

This loop produces the same result as the loop you built in the last section using *goto* and a label. However, the *for* loop is more compact and easier to read and follow in your program.

You may eliminate any of the control statements, but you must provide a semi-colon to let the compiler know that you are not including a part. For example, your program may not need to assign the initial value to the control variable. In such cases, you will not need the initializing statement. In this case, you can write just the semicolon in place of the initializing statement as shown here:

```
int iCount = 0;
for ( ; iCount <= 5; ++iCount)
```

In fact, carried to its extreme, you can write a *for* loop without any of the control statements:

```
for (;;)
```

This loop statement has no initializer, no control expression and no increment. It will execute endlessly unless the statements in the body of the loop provide an exit. Programmers often refer to this construction as a *forever* or *infinite loop*.

Stopping a Loop Before it Finishes

*Sometimes your program may encounter a condition in a loop that makes it impossible to continue performing the statements in the loop. For example, the forever loop—**for (;;)**—will execute endlessly unless you provide some means of **breaking** the loop. To do this, you use the **break** statement, which is the same **break** statement you used in Lesson 20, "Making Decisions Within a C# Program," to jump out of the **switch** statement.*

*The **break** statement also works inside a loop, and ends the loop abruptly. When your program performs a **break** statement from within a **for** loop, program control*

*passes immediately to the first statement after the loop. Your program **does not** test the condition and it **does not** perform the increment step.*

*For example, suppose your program must read a sequence of lines from a file. You expect to read 10 lines in the file and you set up your **for** loop to repeat the steps 10 times. However, if the file contains only seven lines, when your program tries to read the eighth line, the file will return an end-of-file condition, and the data your program reads in the next three iterations will be invalid. In this case, you could use the **break** statement to end the loop early when you encounter the end of the file.*

*The following program, **Break.cpp**, normally would perform the loop statements 10 times. However, it uses a **break** to terminate the loop when the value of **iCount** reaches 7.*

```
namespace nsForLoop
{
    class clsMain
    {
        static public void Main ()
        {
            int iCount;
            for (iCount = 0; iCount < 10; ++iCount)
            {
                Console.WriteLine ("iCount = {0}", iCount);
                if (iCount == 7)
                    break;
            }
        }
    }
}
```

*When you compile and run **Break.cpp**, you can see that the loop only executes seven times.*

Using the *while* Loop to Perform Statements While a Condition is True

The *for* loop will perform a statement or block of statements a specific number of times as determined by the control variable and the three control statements. Normally, the control statements handle the initialization, testing and incrementing of the control variable.

Sometimes, however, you will not know ahead of time how may iterations your loop will perform. The *for* loop is the most common and easiest to use of the C# loop statements, but it is not always suitable for all conditions. In these situations, you can use the *while* loop statement.

In contrast to the *for* loop, the C# *while* loop provides only a test expression. You must initialize the loop control variable before you enter the loop, and modify the variable yourself as part of the statement that forms the body of the loop. The body of the loop may be a simple statement or a compound statement written within braces. The following snippet shows how to write a *while* loop both ways:

```
while (expression)
    statement;
// or
while (expression)
{
    statements;
}
```

The *while* loop will repeat the statement or statements as long as the control expression is *true*. Somewhere in your loop statements, you must perform some action that eventually will make the expression evaluate to *false* or the loop will continue forever. It is important that you understand that your program will evaluate the condition each time it repeats the loop, even the first time. If the condition is *false* when the loop first starts, your program *will not* perform the loop at all.

Unlike C and C++, the control expression in a C# *while* loop must be an expression that returns *true* or *false*. You cannot simply test for a zero or non-zero value. The true/false requirement does slightly limit the use of the *while* statement, but it does put an end to a common programming error: you cannot accidentally make an assignment in the control expression. For example, suppose you mean to write something like *while (x == 10)* but instead you write *while (x = 10)*. The second expression is a valid control expression in C and C++, but means that the test always will be true and the loop will never exit. In C#, however, the second expression would cause the compiler to generate and print an error message.

The following program, *While.cs*, prompts you to enter a letter and then tells you what letter you pressed. It will continue to loop until you type a **Q** in upper or lower case:

```
namespace nsWhileLoop
{
    using System;
    class clsMain
    {
        static public void Main ()
        {
            int ch = ' ';
            while (ch != 'Q')
            {
                Console.Write ("Type a letter and " +
                                "press Enter: ");
                ch = Console.Read();
                string str = Console.ReadLine ();
                Console.WriteLine ("You pressed {0}",
                                    Convert.ToChar(ch));
                ch &= ~' ';
            }
            Console.WriteLine ("Goodbye");
        }
    }
}
```

You should notice that the program assigns the control variable, *ch*, an initial value before entering the *while* loop. The C# compiler will not compile the program if you do not initialize the control variable.

In the body of the *while* loop, you read the console to get a character as you learned in Lesson 4, "Using C# From the Command Line." The call to *Console.Read()* modifies the control variable. After printing the character, you then convert it to uppercase by using a combination of bitwise and arithmetic operators. By testing only for an uppercase letter, you avoid having to make two tests in the loop control expression. Without this line, you would have to write something similar to the following:

```
while ((ch != 'Q') && (ch != 'q'))
```

Converting a Character to Uppercase

In earlier lessons you learned that you can convert any alphabetic character to lowercase by logically ORing the character with a space. Conversely, you can convert any alphabetic character to uppercase by ANDing it with 223, or 0xDF in hexadecimal notation.

*The statement in **While.cs** performs the same operation, but in a slightly more compact manner. It would be a good statement to remember and to place in your bag of programming tricks.*

*The value 0xDF actually is the bitwise inverse of the space character. You can produce this value without having to remember it by writing ~' '. The NOT operator always produces a result of type **int**, and so will have 32 bits in the value. In dealing with ASCII characters, however, you need only be concerned with the first eight bits. The following shows the binary value of a space, and the value return when you apply the NOT operator to a space:*

```
' ' = 00100000       ASCII value of a space
~' ' = 11011111       Hexadecimal DF
```

*Finally, you AND this value with the character to force it to uppercase using the equality operator **&=**.*

*You should note that ANDing the value with 0xDF only works for ASCII characters. In C#, all characters are Unicode and use 16 bits. The method used in **While.cs** will work regardless of the size of a character.*

The *while* loop continues to execute as long as the control expression evaluates to *true*. If the expression is *false* when you first enter the loop, the statements in the body of the loop will not be executed, even once. There are times, however, that you need to assure that the loop will execute at least once, perhaps to initialize some variables to default values.

A variation of the *while* loop, the *do-while* loop, puts the test expression at the bottom of the loop, thus assuring at least one iteration of the loop. You start the loop by writing a *do* statement, then the body of the loop followed by the *while* keyword and the control expression as shown in the following snippet:

```
do
{
    statements;
} while (expression);
```

There are significant differences between the *while* loop and the *do-while* loop that you should know.

❶ The loop starts by using the *do* keyword. The *while* keyword appears at the bottom of the loop.

❷ The line containing the *while* keyword and the control expression ends with a semicolon. In a *while* loop, you do not write a semicolon after the expression.

❸ The statements that form the object of the *do-while* loop must be a compound statement written within a set of braces. This is true even if there is only one statement.

In the *While.cs* program, you could modify the loop to a *do-while* loop to eliminate the need to initialize the control variable *ch*, as shown in the following program, *While2.cs*:

```
namespace nsDoWhileLoop
{
    using System;
    class clsMain
    {
        static public void Main ()
        {
```

```
        int ch;
        do
        {
            Console.Write ("Type a letter " +
                            "and press Enter: ");
            ch = Console.Read();
            string str = Console.ReadLine ();
            Console.WriteLine ("You pressed {0}",
                                Convert.ToChar(ch));
            ch &= ~' ';
        } while (ch != 'Q');
        Console.WriteLine ("Goodbye");
    }
}
}
```

Using the *foreach* Loop

The *for* loop is very commonly used to access the elements in an array. You know the size of an array when you create the array, or you can determine the size by reading the *Array.Length* property. This makes array access a prime candidate for a *for* loop.

Accessing array elements using a *for* loop is so common that C# provides a special loop statement to perform this operation, the *foreach* loop. The C and C++ languages do not contain this form of the loop. In C#, the *foreach* loop iterates through each element of an array. The following snippet shows how to write a *foreach* loop:

```
// Declare and initialize an array of type int
int [] arr = new int [12] = {0, 1, 2, 3, 4, 5,
                             6, 7, 8, 9, 10, 11};

foreach (int x in arr)
{
    // Statements that use x
}
```

The *foreach* loop uses an "iterator" function to retrieve the value of each successive element in the array, then assign that value to the variable you declare in the expression, in this case *x*. There is no control variable or control expression. The variable is in scope only while the loop is executing, and the name of the variable may duplicate the name of another variable in the function.

Note that the iterator function only returns the value of the array elements. You cannot set the values of the array elements using the *foreach* loop. To do this, you need to resort to the trusty *for* loop. The following program, *ForEach.cs*, initializes the elements of an array with a random number using the *for* loop, then prints the element values to the screen using a *foreach* loop:

```
namespace nsForEachLoop
{
    using System;
    class clsMain
    {
        static public void Main ()
        {
            DateTime t = DateTime.Now;
            int n;
            unchecked
            {
                n = (int) (t.Ticks & ((long) ((uint) -1)));
            }
            Random rand = new Random (n);
            int [] arr = new int[20];
            for (int i = 0; i < arr.Length; ++i)
            {
                arr[i] = rand.Next(20);
            }
            n = 0;
            foreach (int i in arr)
            {
                Console.WriteLine ("arr[{0}] = {1}", n, i);
                ++n;
            }
        }
    }
}
```

Notice particularly that, unlike the *for* loop, the variable *i* contains the value of the array element and *not* the index into the array.

Understanding Nested Loops (A Loop Within a Loop)

In Lesson 20, you learned that you can *nest* one decision-making statement within another. This is possible because a decision-making statement is just another statement in C#. The loop statements themselves are simply statements in C#, and you may nest one within another, even if the nested loop is a different loop type from the outer loop. For example, you may nest a *while* loop inside another *while* loop, or a *for* loop inside a *while* loop.

When your program encounters a loop within another loop, it will execute the inner loop through all of its iterations each time the outer loop executes. This will be obvious when you compile and run the following program, *Nested.cs*:

```csharp
namespace nsNested
{
    using System;
    class clsMain
    {
        static public void Main ()
        {
            int ch;
            do
            {
                Console.Write ("Type a letter " +
                                "and press Enter: ");
                ch = Console.Read();
                string str = Console.ReadLine ();
                for (int x = 0; x < 3; ++x)
                {
                    Console.WriteLine ("You pressed {0}",
                                    Convert.ToChar(ch));
                }
                ch &= ~' ';
            } while (ch != 'Q');
            Console.WriteLine ("Goodbye");
        }
    }
}
```

When you run this program, each time you type a character and press **Enter,** the program will execute the inner loop three times, printing the output line each time. Then, the control returns to the outer loop, and you type another character.

WHAT YOU MUST KNOW

In this lesson, you learned how to write loops in your C# code. Looping causes your program to repeat a statement or block of statements based on a control variable and control expression. In C#, the loop comes in four different forms, and you may select the form that best suits your program code. You also learned how to nest loops by placing one loop, the inner loop, inside of another loop. In Lesson 22, "Getting Started with C# Functions," you will learn how to break your program down into small, manageable pieces by placing related code in a function. Before you continue with Lesson 22, however, make sure you have learned the following key concepts:

- Looping is the process of making your program repeat a statement or block of statements.

- Loops will continue to execute while the control expression evaluates to *true*. Your code must modify the control variable so that the loop eventually will end.

- The *for* loop includes statements to initialize, declare and modify the value of the control variable.

- The *while* loop will execute indefinitely as long as the control expression evaluates to *true*. If the expression is *false* when you enter the loop, the loop statements will not execute.

- To assure that your program executes the statements in a *while* loop at least once, you use a variant called the *do-while* loop.

- The *foreach* loop is a special C# loop construct that you use with arrays.

GETTING STARTED WITH C# FUNCTIONS

*i*n previous lessons, you have learned about variables and how a program uses variables to store information. You have used variables to build decision-making statements and to make your programs execute a statement or group of statements repeatedly. Up to now, however, most of your programming has been in the context of the *Main()* function. As your programs increase in complexity, however, you will find it easier to break your code into smaller blocks called *functions*. A function is a named block of code that your program may execute at any time by *calling* the function. In this lesson, you will learn how to write and use functions in a C# program. By the time you finish this lesson, you will understand the following key concepts:

• To define a function, you must give the block of code a name. Your program will use this name to execute the code in the function.

• A function definition must include a *parameter list* even if the function does not use any parameters.

• In C#, all functions must be members of a class or structure. C# does not allow *global* function definitions.

• To call a function, you write the function name and include any necessary values in an *argument list*.

• All functions in C# must have a *return type* even if a function does not return a value.

• C# functions may return one and only one value, but you may use reference arguments if you need the function to change more than one value.

> **Breaking Your Program Into Smaller, More Manageable Pieces**

A *function* is the basic building block of any C# program. A function is a named block of code, usually containing related statements, that performs a specific task. In C#, after you define and write the code for a function, your program may execute the code by *calling* the function.

The C# function is the equivalent of what other programming languages call a *subroutine* or a *procedure*. In object-oriented programming, programmers also refer to functions as *methods*. There are very subtle differences between a function and a method, but for practical purposes the two words are synonymous.

Every C# program must have at least one function, *Main()*, that executes when the program starts. If you limit all of your code to the *Main()* function, however, the function quickly would become long and unmanageable. If all your code is in a single function, making a change to one part of the program might inadvertently cause errors in another statement.

Using multiple functions, you can protect your code against these accidental errors. Changes made to variables in one function do not affect variables in another function unless you specifically provide for that possibility.

To write a function definition, you first write the access keyword such as *public*, *private* or *protected*. Then, you write the function's *return type*. You will look at return types later in this lesson. Next, you write the function's name, which must not begin with a number or a C# operator symbol. Following the function name, inside a set of parentheses, you write the function's *parameter list*, which declares the variables that your program will pass when you call the function. Finally, the body of the function—the actual code that the function will execute when you call it—must be a compound statement, beginning with an open brace and ending with a close brace. The following is an example of a function definition:

```
protected void MyFunc (int x, long y)
{
    // Statements
}
```

You must include the parameter list even if the function does not use parameters. To write a function without parameters, simply write the parentheses together:

```
protected void MyFunc ()
```

In Lesson 18, "Writing Expressions in C#," you wrote a program to accept a value from the keyboard, convert it to a number representing a temperature in degrees Fahrenheit, and then convert the value to degrees Celsius and write the value to the screen. The following program, *FirstFun.cs*, places the conversion and output in a separate function:

```csharp
namespace nsFirstFunction
{
    using System;
    class clsMain
    {
        static public void ShowTemp (int iFahr)
        {
            int iCelsius = 5 * (iFahr - 32) / 9;
            Console.Write("{0} degrees F = {1} degrees C\n",
                          iFahr, iCelsius);
        }

        static public void Main ()
        {
            int iTemp;
            while (true)
            {
                Console.Write ("Enter a temperature " +
                              "in degrees Fahrenheit: ");
                string strTemp = Console.ReadLine();
                try
                {
                    iTemp = Convert.ToInt32(strTemp);
                    ShowTemp (iTemp);
                    return;
                }
                catch (FormatException e)
```

```
        {
            Console.WriteLine (e.Message);
            Console.WriteLine ("Please enter a " +
                                "numeric value");
        }
    }
  }
 }
}
```

The advantage to using the *ShowTemp()* function is that any time you must convert and display a temperature, you do not need to write the conversion formula again, nor must you write the *Console.Write()* function again. You simply call the function and let it perform the work. Using functions reduces the amount of typing you have to do, and thus reduces the chance of program errors.

C# Does Not Allow Global Functions

The C and C++ languages permit you to define functions in *global* space. Global functions are those that are not members of a class or structure. Global space is like a free-fire zone; functions defined in global space may be called from statements anywhere in your code, even other functions that are members of a class.

C#, on the other hand, is a strongly object-oriented language. Functions that you write in your program must be part of an object, such as a class or structure. C# does not let you write functions in global space.

Because all functions are members of a class or structure, you may assign them a protection level of *public*, *protected*, or *private*. This effectively limits the range of statements that may access a member function. You learned about these keywords in Lesson 11, "Understanding Class Scope and Access Control."

You may define functions using the *static* keyword. When you use this keyword along with the *public* access level, you essentially are creating the equivalent of a global function. You have learned that you may call *static* functions without declaring an instance of a class. When you add the *public* keyword, you effectively give the function a global scope.

In the last topic, the *ShowTemp()* function is such a "global" function, as is the standard entry function *Main()*. From within the class, you call *ShowTemp()* simply by writing the name of the function followed by the argument list. If you have other classes in your program, you need only add the class name to reference to the call.

Very often you must provide such utility functions that you may call from anywhere in your code. The C# library functions use this technique. For example, you are able to call the *Console.WriteLine()* function because *WriteLine()* is a *static* member of *Console*.

The following program, *Globs.cs*, declares a utility class *clsTemps* to hold several *static* functions to get temperature values from the console and then to convert the values to or from Celsius and Fahrenheit.

```csharp
namespace nsGlobals
{
    using System;
    class clsMain
    {
        static public void Main ()
        {
            int iTemp = clsTemps.GetFahrTemp ();
            clsTemps.F2C (iTemp);
        }
    }
    class clsTemps
    {
        static public int GetFahrTemp()
        {
            string str = "Enter a temperature " +
                        "in degrees Fahrenheit: ";
            int iTemp = GetTemp (str);
            return (iTemp);
        }
        static public int GetCelsiusTemp()
        {
            string str = "Enter a temperature " +
                        "in degrees Celsius: ";
```

```
        int iTemp = GetTemp (str);
        return (iTemp);
}
static private int GetTemp(string str)
{
        int iTemp;
        while (true)
        {
            Console.Write (str);
            string strTemp = Console.ReadLine ();
            try
            {
                iTemp =Convert.ToInt32(strTemp);
                return (iTemp);
            }
            catch (FormatException e)
            {
                Console.WriteLine (e.Message);
                Console.WriteLine ("Please enter a " +
                                   "numeric value");

            }
        }
}
static public void F2C (int iFahr)
{
        int iCelsius = 5 * (iFahr - 32) / 9;
        Console.Write("{0} degrees F = {1} degrees C\n",
                    iFahr, iCelsius);
}
static public void C2F (int iCelsius)
{
        int iFahr = 9 * iCelsius / 5 + 32;
        Console.Write("{0} degrees C = {1} degrees F\n",
                    iCelsius, iFahr);
}
    }
}
```

The *Globs.cs* program is longer than the *FirstFun.cs* program, but it is much more flexible. You should notice that *Main()* now contains only two lines of code, one to call the function to get a temperature and the other to convert the value and print the new temperature to the console. To change the program from one that converts Fahrenheit to Celsius to convert in the other direction, you need only change the calls in these two lines.

Calling Functions

To execute the code in a function, your program must *call* the function. When you write a function, you declare the *parameters* that it will need using a set of parentheses after the function name. In the parameter list, you specify the data type of each parameter.

When you call the function from within your code, you specify these values as *arguments* to the function. You do not need to specify the data types again in the function call.

For example, the following defines a function that includes a parameter list:

```
void MyFunc(int x, long y, double z)
{
}
```

Because C# does not distinguish between the definition and declaration of a function, you must give identifiers to the parameters if your code uses the parameters. If, however, the function does not use a parameter in the code, you must give it a name:

```
void MyFunc(int x, long, double z)
{
}
```

In this case, the second parameter is simply a "place holder." You still must pass a value of the proper data type in any call to the function, but because the parameter does not have a name, your function may not use it. You may use this technique when you plan to expand the purpose of the function later, and you want to avoid warning messages from the compiler that your code does not use a parameter.

Understanding Function Overloading

*Occasionally, you must write functions to perform similar operations, but using different data types for the parameters. For example, you may want to write a function that returns the square of a **long** number, but later you find that you also must use it to compute the square of a **double** value.*

*You could write separate functions, say **LongSquare()** to return the square of a **long** value, and **DoubleSquare()** to return the square of a **double** value. In C#, you may use the same function name for both functions through a process known as **overloading**.*

*The C# compiler recognizes a function by its **signature**, which is a combination of the function name and the number and data type of the parameters. If you use the same function name, but use different data types for the parameters, the compiler will be able to recognize the functions.*

For example, you might write the following two functions to compute squares:

```csharp
public long Square (long val)
{
    return (val * val);
}
public double Square (double val)
{
    return (val * val);
}
```

Although the function names are the same, the compiler recognizes them as unique functions, and you need not remember the names of two functions. The return type is not a part of the signature. In Lesson 32, "Overloading Functions and Operators," you will learn more about function overloading.

Using Function Return Types

When you define a function, you must give the function a *return type*. This is the data type of the value that the function will return the calling statement. The return type may be a fundamental data type such as *int*, *long*, or *char*. Or, the return type may be a reference to a data type that you have defined, such as a class or structure.

One of the advantages of using functions is that a function definition is like a small program within your program. You can use a function to perform a specific task, then return the result of that task to the statement that called the function. That statement may assign the return value to a variable, which you may use later in your program. Each time your program must perform the task, it only must call the function.

The following program, *Return.cs*, uses several of the constructs you have learned in the last few lessons to accept a numeric score value and convert it to a letter grade:

```
namespace nsReturn
{
    using System;
    class clsMain
    {
        static public void Main ()
        {
            int iScore;
            char chGrade;
            Console.Write ("Enter your score and " +
                           "press enter: ");
            string str = Console.ReadLine();
            try
            {
                iScore = Convert.ToInt32(str);
                chGrade = CalculateGrade (iScore);
            }
            catch (FormatException e)
            {
```

```
            Console.WriteLine (e.Message);
            Console.WriteLine ("Please enter a " +
                                "numeric value");
            return;
        }
        Console.WriteLine("Your grade is {0}", chGrade);
    }

    static char CalculateGrade (int iTestScore)
    {
        char ch;

        if (iTestScore >= 90)
            ch = 'A';
        else if (iTestScore >= 80)
            ch = 'B';
        else if (iTestScore >= 70)
            ch = 'C';
        else if (iTestScore >= 60)
            ch = 'D';
        else
            ch = 'F';
        return (ch);
    }
}
}
```

If you do not return a value from a function, then you must declare the function
with a return type of *void*.

Using the *ref* Keyword to Change Parameters

There are times when a single return value will not be adequate for the task at
hand. For example, suppose you want to calculate the x and y coordinates of a
point on a circle. In this case, you can declare the return type *void* and use *reference* variables in the function call. Reference variables, as you have learned,
allow the function to modify the original values, as shown in the following program, *Point.cs*:

```
namespace nsPoint
{
    using System;
    class clsMain
    {
        static public void Main()
        {
            clsCircle Circle = new clsCircle (300);
            int x = 0;
            int y = 0;
            Circle.Point (42, ref x, ref y);
            Console.WriteLine ("The point on the circle " +
                              "at angle {0} is ({1}, {2})",
                              42, x, y);
        }
    }

    class clsCircle
    {
        public clsCircle (int Radius)
        {
            m_iRadius = Radius;
        }
        private int m_iRadius;
        public void Point (int iAngle, ref int x, ref int y)
        {
            double fAngle = iAngle / 57.29578;
            x = (int) (m_iRadius * Math.Cos (fAngle));
            y = (int) (m_iRadius * Math.Sin (fAngle));
        }
    }
}
```

You should notice two things in *Point.cs*. First, you did not declare the *Point()* function in *clsCircle* as *static*. Therefore, to access the function, you first must declare an instance of the class. That is the first line in the *Main()* function.

Second, although your program will modify the values of *x* and *y*, you still had to initialize the variables before passing them as arguments. The C# compiler will not let you use a variable until you have given it a value. In Chapter 23, you will examine the process of changing parameter values in detail.

WHAT YOU MUST KNOW

In this lesson, you learned how to break your program down into functions, which are smaller, named sections of code. You learned how to declare a parameter list for a function when the function needs additional information to perform its job. Then you learned the basics of how to pass information to those parameters through arguments in a function call. In Lesson 23, "Passing Variables and Values to Functions," you will learn the different methods C# uses to pass those arguments to functions as references or values. Before you continue with Lesson 23, however, make sure you have learned the following key concepts:

- A function is a named block of code that contains related statements.

- You declare and use functions to perform specific tasks. To define a function, you must give the function a return type, a name and a parameter list.

- You must include a parameter list in a function definition even if your program does not require parameters. In this case you provide an empty parameter list by writing the open and close parentheses together.

- All functions must have a return type. If the function does not return a value, you may declare the return type as *void*.

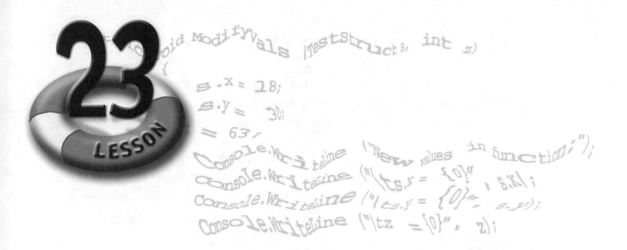

PASSING VARIABLES AND VALUES TO FUNCTIONS

a s you know, as programs execute, they store information in variables. When you create functions, the functions often must store information during the course of their processing. The variables that you declare within a function are *local* to the function, meaning other code that lies outside of the function cannot access the variables and values they contain. When your code calls another function, you often may need to give the other function information that it needs to complete its task. A function that converts a temperature from one system to another must know, for example, the value of the temperature that you want to convert. When you call a function, you specify the values and variables that the function will need as *arguments* to the function. In this lesson, you will learn how to declare and pass arguments from one function to another. By the time you finish this lesson, you will understand the following key concepts:

- Arguments are values, variable and objects that you pass to a function when you call the function.
- C# uses two types of variables, value-type and reference-type. C# stores the values for the two variable types in different ways.
- Your program uses a memory area called the stack to store program information as well as the values for value-type variables.
- For reference-type variables, your program reserves memory in a separate section called the heap. It then stores a pointer to this memory area on the stack.

Understanding Function Arguments and Parameters

In the previous lessons, you have learned how to write functions, and how to *call* those functions. In programming, you often will use functions to perform a specific task, thus breaking your program into small and more manageable pieces.

Often functions that you call will need you to provide information in order to perform their specific tasks. You provide this information by placing *arguments* in an *argument list* as part of the calling process. You specify the function's arguments inside a set of parentheses immediately following the function name. The following statement, for example, passes one argument to a function named FuncName:

```
FuncName(arg1);
```

When you must pass more than one argument to a function, you separate the arguments with a comma. You must specify the arguments in the list in the same order as the parameter list for the function you are calling. You create the parameter list when you define the function.

For example, the *Math.Max()* function tests the relationship between two numbers and returns the value of the number that is greater. To perform the comparison, the function must know the two values that you want to compare. When you call the function, you specify the values as arguments to the function, separated by a comma, as shown here:

```
Math.Max(arg1, arg2);
```

How a C# Program Stores Values

As you have learned, the C# language reference defines two *types* of variables, the *value*-type and the *reference*-type. Your programs store these two types in very different ways.

In programming, you often will see the word *stack* used in relation to variables and function calls. When you load a program into memory, each program sets aside a special section of memory, called the stack, to store program information while the program is running.

The central processor chip in your computer contains circuitry to hold a value that points to a location in the stack. The part of the CPU that holds the value is the *stack register*, and the value that it stores is the *stack pointer*.

When you declare a value-type variable, your program checks the size of the data type of the variable, and then sets aside room to hold the value in the stack. Then, your program adjusts the stack pointer by the size of the data type, so that the pointer again points to the top of the stack. If you declare a variable within the function, your program code (behind the scenes) will store the variable on the stack. When your function completes its processing and returns control to the caller, your program code (again behind the scenes) adjusts the stack pointer to point to the location it was referencing before the function call. As a result, the program essentially discards any information on the stack for the function, which means the value of the function's local variables are lost. Because a function's variables only exist on the stack while the function is active, its variables are said to be *local* to the function and you may not access these variables from outside the function.

For a reference-type variable, your program uses the stack and another portion of memory called the *heap* or *free store*. The heap is a segment of memory set aside for the program to store variables temporarily. In C and C++, the programmer is responsible for maintaining this part of memory. When the programmer reserves a block in the heap, he or she must later free the memory. In C#, the Common Language Runtime manages the heap, which removes the burden from the programmer of having to free memory that the program no longer requires.

When you declare a reference-type variable, such as a variable to hold an instance of a class, your program sets aside memory on the stack to hold the memory location in the heap where it will store the object. Simply declaring a reference-type value does not create the object. To do that you must use the *new* operator. When you use the *new* operator, your program reserves space in

the heap to hold the object, and then places the memory address of that space in the memory that it reserved on the stack.

The program allocates the memory from the heap. However, the program then assigns the memory address of the allocated memory to a variable. As you just learned, a function's variables reside in the stack. Hence, the previous discussion gets a little cloudy as to the use of the stack and the heap. In general, programs will allocate memory from the heap. Function variables, however, will reside in the stack. The reason programmers distinguish between the stack and the heap is because in C#, the Common Language Runtime manages the heap's contents.

When a function ends and returns to the caller, the stack location that contains the heap location is lost, and you no longer can access the object. The object still exists in the heap, but your program no longer knows the object's location in memory. Eventually, the Common Language Runtime will recognize the object is no longer in use and will free the heap object.

Understanding *static* Variables

Your program contains yet another section of memory that it uses to store values that it needs for the life of the program. In other programming languages, this is the *global* data area, but C# scrupulously avoids the use of "global" in describing this memory. When you declare a *static* variable, your program creates the value in this data area, and the value is shared through successive instances of the object containing the *static* variable. Thus, the value is not lost when a function returns. The C# compiler automatically initializes the value of a *static* variable to 0, but you may initialize the variable to another value. When you do assign a value to the variable, your program performs the assignment only once. The following program, *Static.cs*, illustrates the use of a static variable:

```
namespace nsStatic
{
    using System;
    class clsMain
    {
        static int Counter = 1;
```

```
    static public void Main ()
    {
        for (int iCount = 0; iCount < 5; ++iCount)
        {
            clsMain main = new clsMain ();
            main.ShowCounter ();
        }
    }

    private void ShowCounter ()
    {
        Console.WriteLine ("The function has been " +
                            "called {0} times", Counter);
        ++Counter;
    }
}
}
```

Compile and run *Static.cs* and you should see the following output, showing the *static* variable increasing with each call to the function:

```
The function has been called 1 times
The function has been called 2 times
The function has been called 3 times
The function has been called 4 times
The function has been called 5 times
```

Notice that the *static* variable is a member of the *clsMain* class. In the *Main()* function, you declare multiple instances of *clsMain*, but the variable retains its value in each instance. When you increment the value in *ShowCounter()*, the variable retains its new value. Each instance of the *clsMain* class does not re-initialize the *static* variable.

Unlike C and C++, you cannot declare a *static* variable in a function. The reason for this is that all functions must be members of a class, and thus will have access to the *static* members.

Understanding What Happens When Your Program Calls a Function

The stack is more than just a place to store the values of variables. Your program uses the stack for many other purposes, many of which may not be evident.

When you call a function, for example, your program places the memory location of the statement containing the function call in this memory area. When the function returns, the program retrieves this address and returns to the calling function. This is how your program "remembers" the statement where it should resume after a function call.

Before your program performs a function call, however, it first evaluates all the expressions in the argument list, then places the values on the stack. In the case of reference-type values, the code places the address of the objects on the stack. Your program must work with an intermediate module, the Common Language Runtime, and thus the order in which your program stores variables on the stack is important. The C# language specification does not specify this order, and it may be different on other operating systems that implement the Common Language Runtime.

The following diagram, Figure 23.1, gives a general description of how your program would use the stack in a typical function call:

```
FunctionName(x, y);
```

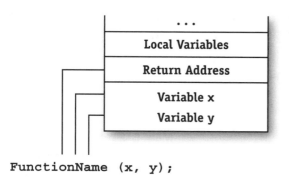

FunctionName (x, y);

Figure 23.1. The C# stack grows upward. Your program stores the values of x and y plus the address of the calling statement on the stack. Then, C# creates any local variables on the stack.

In the case of a reference-type variable, your program reserves space on the stack to hold the address in the heap where the variable is located, as opposed to placing the variable's value on the stack.

A Reference is Like a House Address

When you declare a variable for a reference-type object, you do not actually create the object. Instead, you set aside room on the stack to hold a **pointer** to the object, which you then must create in the heap using the **new** operator.

Think of a pointer to a reference object as being like the address of a house. A friend gives you his address, and you write it down on a piece of paper and place it in your wallet. The piece of paper is the variable you create on the stack, and is not the actual residence where your friend lives. When you must visit your friend, you look at the piece of paper to get the address, and then drive to the house where your friend actually lives.

In C#, when you must use a reference object that you created in a function, your program refers to the value stored on the stack to find out where the object really lives. When your program exits, the value on the stack—the piece of paper in your wallet—gets destroyed and your program no longer knows where the object is located. Eventually, the Common Language Runtime will recognize that your program no longer can access the object and free the memory.

Passing Arguments to a Function

As you have learned, you *call* a function by writing the name of the function, followed by an argument list. The arguments may be values, value-type variables, or reference-type values. A value used as an argument is the result of an expression and C# passes the result on to the called function. However, C# handles value-type and reference-type arguments differently.

In the case of a value-type variable, C# *copies* the value and passes the value as the argument. If the variable is a basic data type, it copies the value that is stored in the variable's memory location. If the argument is a structure, which is a value-type, it passes a copy of the structure to the called argument. The called function, therefore, cannot modify the original value, as shown in the following program, *ValType.cs*:

```
namespace nsValType
{
    using System;
    public struct TestStruct
    {
        public int x;
        public int y;
    };
    class clsMain
    {
        static public void Main ()
        {
            TestStruct s;
            s.x = 12;
            s.y = 24;
            int z = 42;
            ModifyVals (s, z);
            Console.WriteLine ("\nValues after " +
                                "function call:");
            Console.WriteLine ("\ts.x = {0}", s.x);
            Console.WriteLine ("\ts.y = {0}", s.y);
            Console.WriteLine ("\tz = {0}", z);
        }

        static void ModifyVals (TestStruct s, int z)
        {
            s.x = 18;
            s.y = 30;
            z = 63;
            Console.WriteLine ("New values in function:");
            Console.WriteLine ("\ts.x = {0}", s.x);
            Console.WriteLine ("\ts.y = {0}", s.y);
            Console.WriteLine ("\tz = {0}", z);
        }
    }
}
```

When you compile and run *ValType.cs*, you well see the following output, which shows that the function call did not modify the original values:

```
New values in function:
    s.x = 18
    s.y = 30
    z = 63

Values after function call:
    s.x = 12
    s.y = 24
    z = 42
```

Using the *ref* Keyword to Pass Parameters by Reference

Depending on the processing your function performs, there may be times when the function must change the value of one or more parameters. In such cases, you can pass arguments, even value-type arguments, as reference-types, by adding the keyword *ref* in the parameter list for the function definition. You must also precede each argument you want to pass by reference with the *ref* keyword when you call the function. For example, using the previous application, change the *ModifyVals()* function definition for *ValType.cs* as shown in the following:

```
static void ModifyVals(ref TestStruct s, ref int z)
```

Then, in the *Main()* function, change the call to *ModifyVals()* as shown in the following:

```
ModifyVals(ref s, ref z);
```

The *ref* keyword tells the compiler to use the value-type variable as a reference variable. This will allow the function to modify the original values in *Main()*, as shown in the following output:

```
New values in function:
    s.x = 18
    s.y = 30
    z = 63

Values after function call:
    s.x = 18
    s.y = 30
    z = 63
```

C# always passes reference-type variables in this way. You must remember this when you pass an instance of a class as a parameter. If you modify members of the class in the called function, the values of the original members will change.

The ref Keyword Passes an Address

*When you use the **ref** keyword on a value-type argument to a function, you instruct the compiler to use the address of the argument rather than simply to copy the value. By using the address, the function you are calling knows where to find the actual value of the parameter. To complete the call, you must be sure the value you pass is a reference. If you try to pass a value to a function that expects a reference, the compiler will issue an error.*

*For example, in the following program, **RefType.cs**, the definition for the **SquareIt()** function states that the function expects a reference to an **int** value:*

```
namespace nsRefType
{
    using System;
    class clsMain
    {
        static public void Main ()
        {
            int x = 4;
            NoSquare (x);
            Console.WriteLine ("x after NoSquare = " + x);
```

```
      SquareIt (ref x);
      Console.WriteLine ("x after SquareIt = " + x);
    }

    static protected void NoSquare (int square)
    {
      square *= square;
    }

    static protected void SquareIt (ref int square)
    {
      square *= square;
    }
  }
}
```

*Although the statements in the two functions perform the same operation, **NoSquare()** takes only a value as its parameter, and so it cannot change the original value. However, you pass **SquareIt()** a reference by using the **ref** keyword both in the function definition and in the statement that calls the function. By using a reference, **SquareIt()** changes the original value of **x**.*

WHAT YOU MUST KNOW

In this lesson, you learned how C# stores information in variables, and how your program uses different parts of program memory to store variables of different types. You also learned how your program handles variables in a function call, and how to pass variables of different types in a function call. In Lesson 24, "Using Objects, Values and Constants as Arguments," you will learn how to cast variables to different types, how to use arrays as arguments and how to use "constant variables." Before you continue with Lesson 24, however, make sure you have learned the following key concepts:

- A C# program stores variables in three different parts of memory, the stack, the heap and global memory.

- How you declare a variable determines what section of memory your program uses to store the variable's value.

- Your program always stores reference-type variables in the heap. However, it stores a pointer to the storage on the stack or in global memory. Ordinarily, you cannot access the pointer.

- When you pass value-type variables to a function, the called function cannot change the value of the original variable.

- You can permit the called function to modify the original values by passing the variables as reference-types using the *ref* keyword.

USING OBJECTS, VALUES AND CONSTANTS AS ARGUMENTS

*f*unctions often need information to perform their tasks, and you provide this information by passing arguments to the functions. C# uses value-type and reference-type variables, and uses these types in different ways when you pass them as arguments. C# is a strongly typed language, and the compiler enforces type checking. It will prevent you from assigning the wrong data type to a variable, for example. There are times, though, when you will want to override this behavior. In this lesson, you will learn how to use arrays and structures as arguments and how to override the data type of a variable. You also will learn how to use "constant variables" to protect reference objects from accidental modification. By the time you finish this lesson, you will understand the following key concepts:

- You may pass values, including objects such as an instance of a class or structure, to a function as an *argument*.
- When you call a function using a value-type variable, C# makes a copy of the value. The function may modify the value, but the changes do not affect the value of the original variable.
- When you call a function using a reference-type variable, C# passes the *address* of the variable to the function. Changes you make in the function will affect the original variable.
- You may change the data type of a variable temporarily by using the *cast* operator.

- A constant is a variable that your program may assign a value to only once. Afterward, your program may not modify the value.
- You must give a constant a value when you declare the constant.

Using Structures and Arrays as Arguments

As you have learned, an *array* is a data structure that lets you store multiple values of the *same* data type using only one variable name. The data type may be any of the fundamental data types such as *int* or *double*, but all the elements of an array must be of the same type. When you declare and create an array, you actually are creating an instance of the *System.Array* class. Variables that hold arrays, then, are always reference-type variables.

When you use an array as an argument to a function call, your program passes a reference to the array. Thus, if you make any changes to the array in the function you call, you will modify the values in the original array.

The following program, *ArrArg.cs*, shows that changing the values of an array during a function call will change the original array values:

```
namespace nsValType
{
    using System;
    class clsMain
    {
        static public void Main ()
        {
            int [] arr = new int [] {1, 2, 3};
            Console.WriteLine ("Before function call:");
            Console.Write ("\tThe array contains " +
                        "{0}, {1}, {2}\n",
                        arr[0], arr[1], arr[2]);
            ModifyVals (arr);
            Console.WriteLine ("After function call:");
            Console.Write ("\tThe array contains " +
                        "{0}, {1}, {2}\n",
                        arr[0], arr[1], arr[2]);
        }
```

```
static void ModifyVals (int [] arr)
{
    for (int x = 0; x < arr.Length; ++x)
        arr[x] *= 2;
    Console.WriteLine ("In function call:");
    Console.Write ("\tThe array contains " +
                   "{0}, {1}, {2}\n",
                   arr[0], arr[1], arr[2]);
    }
  }
}
```

A *structure* on the other hand, is a data structure that lets you store multiple values, even values of different types, using only one variable name. A structure is very similar in design to a class, but very different in the way C# uses structures and classes. A variable holding a structure normally is a value-type variable.

If you pass a structure as an argument in a function call, your program will copy the original structure and pass the copy as the argument. Changing the values in a structure in a function call will not change the original values.

Temporarily Changing the Data Type of a Value

As you have learned, C# is a strongly typed language, which means that the compiler performs type checking to make sure the data types of the variables you use in function calls and expressions are the same. If the data types do not match, the compiler will automatically promote a data type of a lower precision to one of a higher precision.

When the compiler performs this type of automatic conversion from one data type to another, it is an *implicit* conversion. There are times when implicit conversion will not work, however, and you must temporarily change the data type of a variable through *casting*. When you cast one data type to another in your code, you are performing an *explicit* conversion. You must be aware of the difference because, when you mix data types in a program and the compiler issues an error, the compiler will use these words in the error message.

For example, you have learned that the *Console.Read()* function returns a character from the keyboard as a value of type *int*. If you later try to use that *int* value in a function call that requires a type *char*, the compiler will reject the function call and report an error. The following snippet shows an example of this:

```
static public void Main()
{
    int x = Console.Read();
    CheckChar (x);
}
static public void CheckChar (char ch)
{
}
```

If you add this code to a program and try to compile the program, the compiler will issue an error message similar to the following:

```
Args.cs(19,22): error CS0029: Cannot implicitly convert type
'int' to 'char'
```

The compiler is telling you that you cannot use the variable *x* as an *int* when you pass it as an argument to *CheckChar()*. An *int* type variable uses 32 bits to store its value, and a *char* type uses only 16 bits, so the compiler rejects the implicit conversion because you would lose precision. However, you know that *x* really contains a *char* value and the conversion would not cause a loss of precision. In this case you must do an explicit conversion in the function call. To perform an explicit conversion, you write the new *temporary* data type inside a set of parentheses before the variable name:

```
CheckChar((char) x);
```

An explicit conversion is temporary, and applies only to the one instance in which you cast the value from one data type to another. In later statements, *x* still is of data type *int*. If you must use *x* repeatedly as a type *char*, you could perform the cast on the *Console.Read()* return value:

```
char x = (char) Console.Read();
CheckChar(x);
```

Your program then will perform the explicit conversion before it assigns the character to *x*.

You cannot do an explicit conversion if the data types are not compatible. For example, you cannot explicitly convert an *int*-type variable to a structure-type variable, as in the following snippet:

```
struct TestStruct
{
    public int a;
    public char ch;
}
static public void Main()
{
    TestStruct x = (TestStruct) Console.Read();
}
```

The compiler will reject this cast because the data types are not compatible. Your program stores the values of the two data types in different ways.

Using References

Computer programming languages that enable function or procedure calls use one of two methods to pass variables from the calling statement to the function. In the first method, *call by value*, the program makes a copy of the value of the variable to the function. The program may use and modify the value, but the changes do not apply to the original variable.

In the second method, the program passes the memory location of the variable. This method, *call by reference*, allows functions to modify the value of the original variable directly. Many early computer languages such as BASIC and FORTRAN used the call-by-reference method.

There are advantages and disadvantages to both methods. The C language uses call by value, but lets programmers simulate a call by reference through the use of *pointers*. Using pointers requires the programmer to use a special operator to obtain the address of a variable. Then, to assign a value to the original variable, the programmer must *dereference* the pointer. It is difficult for many beginning programmers to understand and use pointers, but use of pointers is a powerful technique when used by experienced programmers. Pointers make it easier to pass large objects such as structures or arrays to functions.

The C++ language simplified the use of pointers by implementing a *reference variable* type. A reference variable always contains the address of another variable. When you modify a reference variable, the changes actually are made to the variable that is being referenced. Reference variables greatly simplify pointer use, but require the programmer to declare special variables using a reference operator.

C# implements both methods, and in most cases the difference is transparent to the programmer. You have learned that C# defines two types of variables, *value-types* and *reference-types*. C# passes value-types using the call-by-value method, and reference-types using the call-by-reference method.

You may force the compiler to use a value-type variable as a reference-type by using the *ref* keyword. Using the *ref* keyword is similar to casting a variable to a different type, and its use is only temporary. It does not actually change the variable from a value-type to a reference-type.

The *ValType.cs* program in Lesson 23, "Passing Variables and Values to Functions," demonstrates how to force the compiler to use a structure argument as a reference-type rather than as a value-type.

Understanding Constants

In C and C++, you may *define* symbols to represent values. When you compile your program, the compiler's preprocessor scans your program and replaces the symbol with the value it represents. Once you have entered the definition in your program, there is no chance of mistyping the value.

If you deal with graphics, you often will need to use constants to calculate values or to convert from one system to another. For example, you will use *pi*, 3.14159, to calculate points on an arc or circle, and a *radian*, 57.29578 degrees, to convert from angles to the radian system used by most library code. For example, in C and C++ you could use the #define statement to create a constant:

```
#define     RADIAN          57.29578
```

From that point, you could write the identifier *RADIAN* in your code and the preprocessor would replace it with *57.29578*.

C# also implements a preprocessor, but in a very different way. The C# preprocessor lets you define a symbol, but not to assign a value to that definition. Thus, no replacement takes place during the preprocessor stage. Instead, in C# you may use *constant variables* to implement the same functionality.

The concept of a "constant variable" may sound oxymoronic. As you will see, it is an apt description of how you define and use such variables. Programmers often refer to constant variables simply as "constants."

Constants give your program the ability to define a value that your program cannot change accidentally, or even on purpose.

Defining Constants

To use a constant, you first must declare a variable to hold the value. You declare the variable as you would any other variable, but you include the keyword *const* in the definition, as in the following:

```
const double Radian = 57.29578;
```

You must remember that you must assign a constant a value at the time you declare the variable. The following syntax would cause the compiler to issue an error message telling you that a *const* field requires a value:

```
const double Radian;
Radian = 57.29578;
```

The reason for using a constant is to provide a value that will not change. Once defined, you cannot change the value of a constant. The following snippet also would cause the compiler to issue an error message:

```
const double Radian = 57.29578;
Radian = 3.14159;
```

If you must use a different value for a constant, you must declare another *const* variable:

```
const double Radian = 57.29578;
const double Pi = 3.14159;
```

There are a few rules you must understand when using constants. Constants have *scope*, and you must refer to them using the namespace and class in which you define the constant. If you define a constant in a function, then it is in scope only in the function, and you may not use it in another function. The following program, *Const.cs*, shows a constant, *Radian*, as a class member and another constant, *Pi*, defined in a function:

```
namespace nsConst
{
    using System;
    class clsMain
    {
//
// Radian may be used as a static field in clsMain
//
        public const double Radian = 57.29578;
        static public void Main ()
        {
            double Degrees = 60;
            double Radius = 30;
            double Radians = 1.4563;
            Console.WriteLine ("{0} degrees = {1} radian",
                    Degrees,
                    nsMath.clsMath.ToRadians (Degrees));
            Console.WriteLine ("{0} Radians = {1} Degrees",
                    Radians,
                    nsMath.clsMath.ToDegrees (Radians));
            Console.WriteLine ("The area of a circle " +
                    "with a radius of {0} is {1}",
                    Radius,
                    nsMath.clsMath.AreaOfCircle (Radius));
        }
    }
}

namespace nsMath
{
    class clsMath
    {
        static public double ToRadians (double Degrees)
        {
            return (Degrees / nsConst.clsMain.Radian);
        }
        static public double ToDegrees (double Radians)
        {
            return (Radians * nsConst.clsMain.Radian);
        }
        static public double AreaOfCircle (double Radius)
```

```
        {
//
// Pi may be used only in the AreaOfCircle() function
//
            consu double Pi = 3.14159;
            return (Pi * Radius * Radius);
        }
      }
}
```

You should notice that *Radian* is a constant field in *clsMain*. You did not declare *Radian* as *static*, but you did not need to create an instance of *clsMain* to use the constant in your code. Constants that you declare as field members of a class automatically are *static*, and you need not use the *static* keyword on them. In fact, the compiler will generate an error if you add the *static* keyword to the definition.

By contrast, you declared *Pi* in the *AreaOfCircle()* function, so it is a *local* variable. You may use local variables only in the function in which you declare them.

You may use other constants to define a constant, but you may not use a non-static or non-constant variable in a constant definition, as shown in the following examples:

```
const int x = 4;      // Define one constant
const int y = 9 * x;  // The compiler knows the value of x
int z = 3;            // An ordinary field
const a = z * 8;      // Error. z does not exist until you
                      // declare an instance of the class
```

In the preceding example, you declare and initialize the field *z*, but it does not exist until you declare an instance of the class. Because it is not a constant, you may change the value and the compiler has no way of knowing its value when you compile your program.

Using Constants as Arguments

Under certain conditions, you may pass constants as arguments to functions. However, C# does not support the use of the *const* keyword when declaring parameters to functions, as do C and C++. Thus, if you pass a constant as an argument, you should be aware that the called function may modify the local value for its own use, as shown in the following snippet:

```
static public double VolumeOfSphere (double Radius)
{
//
// Pi may be used only in the AreaOfCircle() function, so
// you will need to pass the value as an argument
//
    const double Pi = 3.14159;
    double Area = AreaOfCircle (Radius, Pi);
    return (4 * Area * Radius / 3);
}
static public double AreaOfCircle (double Radius, double Pi)
{
// Ooops. Pi is not a constant and you may get
// erroneous results
    Pi *= 2.9;
    return (Pi * Radius * Radius);
}
```

Changing the value within the called function does not modify the original value in the calling function. You may not pass a constant as a reference because the called function then would be able to modify the value. The following definition of *AreaOfCircle()* is valid:

```
static public double AreaOfCircle (double Radius,
                                   ref double Pi)
```

However, if you try to call the function using a constant as a reference as in the following, the compiler will issue an error that the argument must be an *lValue*:

```
double Area = AreaOfCircle (Radius, ref Pi);
```

WHAT YOU MUST KNOW

In this lesson, you learned how to pass arguments to functions both as values and as references. You learned how C# handles value-type variables and reference-type variables in a function call, and how to beware that you do not accidentally change the value of a reference-type variable. You learned about "constant variables"—or simply "constants"—and how to declare and use constants in your program, both as field members in a class definition and as local variables in functions. In Lesson 25, "Understanding Function and Variable Scope," you will learn about the *scope* of variables. Before you continue with Lesson 25, however, make sure you have learned the following key concepts:

- Computer programming languages may use *call by value* or *call by reference* systems when passing values to functions. C# uses a combination of the two systems.

- When you pass a value-type variable to a function, your program makes a copy of the value. You may alter the value in the called function without affecting the original variable.

- When you call a function using a reference-type variable, C# actually passes the address of the variable to the called function. If you change the value of the variable, the value of the original variable will change.

- Constants are variable declarations that you may not modify once you assign values to the constants.

- You may pass constants to a function, but the called function will not treat the value as a constant. You may not pass constants to a function as a reference argument.

UNDERSTANDING FUNCTION AND VARIABLE SCOPE

*a*s you have learned, by using functions, you may break your code down into smaller, more manageable blocks of code. A function is a named block of code that your program may execute simply by calling the function. In Lesson 24, "Changing Parameter Values Within a Function," you learned how to pass values and references to a function to give the function information it needs to perform a particular task. In addition to the arguments you pass to a function, the function may need other information, such as values in the class definition or in other classes. A function may declare its own variables to store information temporarily. The extent to which a variable may be used is part of the *scope* of the variable. In this lesson, you will learn about scope. By the time you finish this lesson, you will understand the following key concepts:

- The *scope* of a variable is the range of statements that may access the variable. Within its scope, the *fully qualified name* of the variable must be unique.
- Functions as well as variables have scope. Within the scope of a function, you may duplicate the name of a function provided each function has a unique *signature*.
- Variables you declare within a function have *function* scope. You may use the variables only within the function in which you declare the variables.
- You may declare *blocks* within a function and declare variables within the block. Only statements within the block may access variables you declare within a block.
- A variable in a function is visible to your program only after you declare the variable.

- Some variables, especially objects, may persist even after they go out of scope or are no longer visible.

Understanding Scope

Depending upon how and where you declare variables, only a limited number of statements in your program may use the variables you declare. This range of statements is the *scope* of a variable. In Lesson 23, "Passing Variables and Values to Functions," you learned that when you declare variables in a function, your program sets aside space on the stack to hold the value. Even if you create an object in a function, your program sets aside space on the stack to hold a reference, or pointer, to the object.

The variables that you declare within a function do not exist until you call the function and your program begins executing the block of code in the function. Your program creates variables on the stack and adjusts an internal memory location, the *stack pointer*, the code uses to reference the variables. When your function ends, your code adjusts the stack pointer back to the same condition it had when you called the function. Adjusting the stack effectively destroys any variables that you declare within the function and the values are lost.

Variables that you declare within a function are *local* variables, and you may use the values of these variables only within the function. Other functions are not even aware that the local variables exist. The range of statements that may access a variable is the variable's *scope*.

Scope is a very important concept in object-orient programming. Within the scope of a variable, the name—or variable identifier—must be unique. If you also program in C or C++, you must understand that many aspects of scope in C# are very different.

In C#, the name of a variable that you declare in a function is simply the identifier that you give the variable. By contrast, the name of a variable that you declare as a member of a class consists of the namespace, the class name and the identifier. In the following declarations, the two variables have distinct names, although the identifier is the same:

```
namespace nsTest
{
    class clsMain
    {
        static public int iVar;

        static public void Main ()
        {
            int iVar;
        }
    }
}
```

The full name of the variable *iVar* that you declare as a field in the class is *nsTest.clsMain.iVar*, while the full name of the variable in the *Main()* function is simply *iVar*. The sequence of using the namespace, class name, and identifier is the *fully qualified name* of the variable.

In the preceding snippet, the variable *iVar* in the *Main()* function *hides* the class member *iVar*. If you simply refer to the variable as *iVar*, you will access the function variable. To access the class member, you must use as much of the fully qualified name as is needed to uniquely identify the class member variable.

Function Declaration Scope

Functions as well as variables have scope, but the identity of a function goes beyond just the name of a function. The C# compiler identifies a function by its *signature*, and within its scope a function signature must be unique.

The C# compiler builds a signature for a function by using its identifier—the function name—and then combining the fully qualified name with the number and data types of the function's parameters. The actual names you give the parameters are not a part of the signature. If the compiler can uniquely identify a function from the signature, it will allow the function name.

The following program, *FunScope.cs*, declares five functions using the name *TestFunc()*. Not all the declarations are unique.

```
namespace nsOne
{
    class clsMain
    {
        static public void Main()
        {
            // Some statements
        }
// Declaration 1.
// The compiler will accept the first declaration.
        void TestFunc (int x, long y)
        {
            // Some statements
        }
//
// Declaration 2.
// The number of parameters is different than the
// preceding, so the compiler will accept the
// following declaration.
        void TestFunc (int x)
        {
            // Some statements
        }
//
// Declaration 3.
// The data type of parameters is different than the
// preceding, so the compiler will accept the
// following declaration.
        void TestFunc (long x, int y)
        {
            // Some statements
        }
//
// Declaration 4.
// The following declaration differs from the first
// declaration only in the return type. The compiler
// will reject this declaration.
        int TestFunc (int x, long y)
```

```
            {
                return (x * (int) y);
            }
        }
}
namespace nsTwo
{
    class clsClass
    {
//
// Declaration 5.
// This declaration is in a different namespace and
// class, so the compiler will accept the following
// declaration.
        void TestFunc (int x, long y)
        {
            // Some statements
        }
    }
}
```

Notice that the fourth declaration differs from the first declaration only by the return type. The C# compiler does not use the return type as part of the function's signature, so the signatures of the two functions are identical. The compiler will issue an error message.

Writing functions using the same name but with different parameters is a process called *overloading*. You will learn more about overloading in Lesson 32, "Overloading Functions and Operators."

Determining Function Scope

When you declare a variable as a parameter to a function or method, the scope of the variable is the function name declaration plus the following block of code that forms the body of the function. The parameter has *function scope*.

Function scope begins at the opening parenthesis of the parameter list and continues to the closing brace that ends the body of the function. It does not begin

with the function declaration. The difference is subtle, but important, because it allows you to have a variable within a function that has the same name as the function. The following is legal in C# because the function name and the variable name have different scope:

```
void TestFunc ()
{
    int TestFunc = 42;
    Console.WriteLine ("TestFunc = " + TestFunc);
}
```

You probably would not want to write a function this way on purpose, but you should be aware that the construction is valid in C#. In a *recursive* function, the variable name would hide the function name, and would cause the compiler to reject the construction. (A recursive function is one that calls itself; such routines are common in sorting algorithms.) You still could write a recursive function, but you would have to prefix the recursive call with the *this* identifier, as in the following program, *Recurse.cs*:

```
namespace nsRecurse
{
    using System;
    class clsMain
    {
        static public void Main()
        {
            clsMain main = new clsMain();
            main.TestFunc (10);
        }
        int TestFunc (int TestFunc)
        {
            Console.WriteLine ("TestFunc = " + TestFunc);
            —TestFunc;
            if (TestFunc > 0)
            {
                this.TestFunc (TestFunc);
            }
```

```
            return (TestFunc);
        }
    }
}
```

Recurse.cs will print the value of *TestFunc* from 10 down to 1, and then will exit. The function name and the parameter name are the same, but because they have different scope, they do not interfere with one another. In C#, function scope is the same as in C++, but not in C (you could not write such a program in C).

C# Classes are Self-Aware

In the early days of modern computing, before computers became household appliances, Xerox was developing an early version of an object-oriented programming language named "Smalltalk" at its Palo Alto Research Center in California. In Smalltalk, objects have a strange property called "self." Each object that you create in Smalltalk inherits its own copy of the **self** property, which identified the object itself. Thus, an object could distinguish between itself and any other object. The objects were "self-aware."

The C++ programming language carried this concept into classes and structures. Each instance of a class or a structure inherited a pointer called **this** that the object could use to reference itself. C# does not use pointers, other than in **unsafe** code, but it does implement a reference variable that refers to itself. You can use this reference variable by using the keyword **this**. The following program, **Self.cs**, declares two instances of a class, and then calls a class member function, **GetInstance()**, which returns a **this** as a reference to itself:

```
namespace nsSelf
{
    using System;
    class clsMain
    {
        static public void Main ()
        {
//
```

```
// Declare two instances of clsSelf
        clsSelf First = new clsSelf (42);
        clsSelf Second = new clsSelf (64);
//
// Declare a variable to hold a class reference
        clsSelf Instance;
        Instance = First.GetInstance();
        Console.WriteLine ("In first instance, val = "
                              + Instance.val);
        Instance = Second.GetInstance();
        x = Instance.val;
        Console.WriteLine ("In second instance, val = "
                              + Instance.val);
    }
  }
  class clsSelf
  {
      public clsSelf (int value)
      {
          val = value;
      }
      public int val;
      public clsSelf GetInstance ()
      {
          return (this);
      }
  }
}
```

Although you use just one variable, **Instance**, your program can obtain references to each class instance by calling the **GetInstance()** function. When you run **Self.cs**, you should see the following output:

```
In first instance, val = 42
In second instance, val = 64
```

You cannot use the **this** keyword to reference **static** class members, however. In C#, remember, a **static** class member belongs to the class itself, so you must access them using the class name rather than the instance name. The following snippet would cause an error in C#:

```
class clsMain
{
    static public void Main()
    {
        clsSelf First = new clsSelf()
        clsSelf Instance = First;
        Console.WriteLine ("val = " + Instance.val);
    }
}
class clsSelf
{
    static int val = 42;
}
```

To reference the **static** variable, you would need to refer to it as **clsSelf.val**, not by using the instance name.

Function scope allows you to duplicate the names of parameters and variables in other functions. You can have a variable called *iVar* in one function, and a variable of the same name in another function and they will not conflict with each other, as in the following snippet:

```
static public void Main ()
{
    int iVar;
}
static public void AnyFunc()
{
    int iVar;
}
```

The two *iVar* declarations do not interfere with each other because the code in one function cannot access the variable in the other function. When a variable is not visible from a statement being executed, then it is *out of scope*.

You have learned that within the same function variable names must be unique. As with most rules, there is an exception: Variables used in *block* scope may duplicate one another.

Using Recursion on Math Problems

*Recursion is handy for sorting arrays. In C#, however, the **Array** class contains its own sorting method, so you usually will not have to write your own sort routine. But recursion also is a good method for solving many mathematical problems, such as a series, some transcendental numbers, or a factorial. Factorials appear very often in mathematical problems, such as calculating the odds of winning a lottery. The following program, **Factor.cs**, uses recursion to compute the factorial of a number entered from the keyboard:*

```
namespace nsFactorial
{
    using System;
    public class clsMain
    {
        static public void Main (string [] args)
        {
            if (args.Length == 0)
            {
                Console.WriteLine ("Please enter a number");
                return;
            }
            uint Val;
            try
            {
                Val = Convert.ToUInt32(args[0]);
            }
            catch (FormatException)
            {
                Console.WriteLine ("Please enter a number");
```

```
            return;
        }
        if (Val == 0)
        {
            Console.WriteLine ("0! = 1");
            return;
        }
        Console.Write (Val + "! = ");
        uint factorial = Factorial (Val);
        Console.WriteLine (" = " + factorial);
    }
    static public uint Factorial (uint Num)
    {
        Console.Write (Num);
        if (Num > 1)
        {
            Console.Write (" x ");
            return (Num * Factorial (Num - 1));
        }
        return (1);
    }
  }
}
```

When you compile and run **Factor.cs**, run the program with a number from the command line as shown in the following. The program will respond by printing the string of numbers used in calculating the factorial. The **Factorial()** function recursively calls itself, decreasing the value of the argument by *1* on each call, until the argument finally reaches *1*.

```
C:>factor 10   <enter>
10! = 10 x 9 x 8 x 7 x 6 x 5 x 4 x 3 x 2 x 1 = 3628800
```

Note that 0! by definition is 1. To eliminate printing the string of numbers, remove the **Console.Write()** statements in the **Factorial()** function.

Using Block Scope

Within a function, you may break your code into separate *blocks* by writing the code within a set of braces. The block limits the scope of the variables you declare within the block. Within different blocks, you may declare variables using the same name just as though the variables were declared in different functions. The following snippet shows an example of block scope:

```
void TestFunc ()
{
// The braces limit the scope of the variable
    {
        int x = 42;
        Console.WriteLine ("x = " + x);
    }
// The variable x below is in a different scope from the
// preceding declaration.
    {
        int x = 24;
        Console.WriteLine ("x = " + x);
    }
}
```

There are some differences between block scope in C# and C++. In C#, you may not declare a variable in block scope that you have declared in function scope. The following snippet is legal in C++:

```
void TestFunc ()
{
// This x has function scope
    int x = 42;
    {
// This x has block scope
        int x = 24;
    }
}
```

In C++, the variable *x* in the inner set of braces hides the *x* outside the inner braces. The variables have different scope and thus do not interfere with one another. The C# compiler, however, will reject this construction because the inner *x* duplicates the outer *x*.

In a compound statement, the block scope begins at the opening brace and ends at the closing brace. For a conditional or loop statement, the scope begins with the opening parenthesis. In the following code, the two variables named *x* have different scope and are acceptable to the C# compiler:

```
void TestFunc ()
{
// This x has block scope
    for (int x = 0; x < 10; ++x)
    {
        // Statements in the first for loop
    }
// . . .
    for (int x = 0; x < 42; ++x)
    {
        // Statements in the second for loop
    }
}
```

Notice that each *for* loop declares a variable *x*. Because the variables are in different block scopes, the declarations are proper. Variables declared in block scope may not be used outside the block in which you declare the variables. Each block is like a function within a function, and when the block exits, the variables you declare in the block are destroyed. The following would cause the compiler to generate errors:

```
void TestFunc ()
{
// This x has block scope
    for (int x = 0; x < 10; ++x)
    {
        int y = 2 * x;
```

```
        Console.WriteLine ("y = " + y);
    }

// Error. The y above is out of scope and cannot be used
// in the following statement
    Console.WriteLine ("y = " + y);

    for (int x = 0; x < 12; ++x)
    {
        // Statements in the for loop
    }
// Error. x was declared in block scope and cannot be used
// in the following statement.
    if (x == 12)
    {
        Console.WriteLine ("Object not found");
    }
}
```

The difference is important, because testing the loop variable after the loop exits is common practice to determine whether a search for a value in a loop succeeded, such as the *for* loop in the preceding snippet. To do this in C#, you would have to declare the loop variable outside the loop statement:

```
int x;
for (x = 0; x < 12; ++x)
```

Understanding Visibility and Duration

Two properties of scope that you will see from time to time are a variable's *visibility* and *duration*. The visibility of a variable is the range of statements that may access and use a variable's value.

You have learned that you may declare a variable within a block of code. The scope of such a variable is block scope, the block of code in which you declare the variable. However, the variable is visible only to statements following the statement in which you declare the variable. Because a variable is not visible before you declare it, you may not use the variable in an expression in a preceding statement. The following snippet declares a variable *x* inside a block but after the first statement in the block:

```
void TestFunc()
{
    // Statements
//
// Start a block of code
    {
        Statement1;
        int x;
        Statement2;
        Statement3;
    }
}
```

The variable *x* is not visible until the point of declaration, so *Statement1* is never aware that *x* exists, and it may not reference *x*. However, *x* is visible to *Statement2* and *Statement3* and may use the variable.

The *duration* of a variable is the time between the point where your program creates the variable and where your program destroys the variable. For variables that you declare in a function, the duration is the time that your program is executing statements in the function. In the above example, your program destroys *x* at the end of the block.

When you create an object using the *new* operator, the object itself may endure for some time after it goes out of scope or no longer is visible. The Common Language Runtime eventually will destroy the object.

In this lesson, you learned about the scope of variables and functions. You learned what statements in your program can access a certain variable, and that how and when you declare a variable determines the scope of the variable. You learned about block scope, and the visibility and duration of variables. In the previous lessons, you have learned most of the basics of C# programming. In Lesson 26, "Writing a Windows Program," you will begin writing programs for Windows from within the Visual Studio. Before you continue with Lesson 26, however, make sure you have learned the following key concepts:

- How and where you declare a variable directly affects its scope and the statements that may access the variable.

- The C# compiler creates a *signature* for functions using the function name and the number and data type of parameters. If two functions within the same scope have the same signature, the compiler will issue an error.

- Variables declared within a function must have unique names and may be accessed only within the function. Function scope begins with the opening parenthesis after the function name.

- Variables declared within a block may not *hide* other variables declared within the function. However, variables in one block may duplicate the names of variables in another block.

- Variables with block scope may not be used outside the block. Your program destroys the variables with the block ends.

- In a loop statement, the block scope begins at the opening parenthesis of the loop control statement. Thus, if you declare a variable in the loop control statement, you may not use it outside the loop.

WRITING A WINDOWS PROGRAM

C# is like most programming languages in one respect: You will be learning about it for a very long time. Now that you have learned the basics of C# programming, it is time to turn your attention from command-line programs to Windows programs. Windows programming, after all, is the ultimate purpose of the Visual Studio. Most programmers consider writing programs for Windows considerably more involved than writing programs for the command line, but the Visual Studio contains a number of tools to ease that burden. In this lesson, you will write a Windows program that displays a form showing the time. This lesson will prepare you for upcoming lessons that use more complicated forms. By the time you finish this lesson, you will understand the following key concepts:

- Windows applications use the Graphic User Interface to display an application's window and the various *controls* on the window.
- Windows communicates with your application through a series of messages.
- In C#, Windows applications do not use the *Console* class to read from the keyboard or to write to your screen.
- The Visual Studio contains a number of tools and tool windows to help you to write Windows applications.
- A *form* is a dialog box that contains other graphical objects. You use these objects to accept input from the user or to display information to the user.
- When you create a C# Windows application, the Visual Studio always creates a *form* for your project.

Understanding How Windows Programs Differ

Most of your programs to this point have been console programs. You have written them in a text editor such as *NotePad* and then compiled and run the programs from the Windows console. Command-line programs are good for experimenting and learning about the syntax of a language. In addition, if, and when, other operating systems implement versions of C#, command-line programs will be more portable than those that rely on a particular operating system.

Eventually, however, you must experiment with Windows programming. That is, in the final analysis, the underlying purpose of the Visual Studio.

Windows programming is considerably more complex than command-line programming. You must add several additional references to your program. Fortunately, the Visual Studio tools will handle most of the details for you.

Windows programs do not use the *Console* class input and output functions. Instead, you write your text to Windows objects. These objects are *controls* such as a text box, a list box or a static text field. The Visual Studio will handle the details of creating the window, or form, for your project. You must add the controls to the form to which your program will write information.

Windows programs use the *Graphic User Interface*, which you often will see in the MSDN help file as "GUI" (pronounced "gooey"). The GUI treats everything, even text, as graphics. Your program must draw its main window, usually a form in a C# Windows program. Then, it must draw all the controls for the window. To display text, your program creates a *font* and draws the text using the font.

Rather than write or draw directly to your screen, Windows communicates with your program through a series of *messages*. These are not messages in the sense of e-mail, where you can display and read them. Instead, the messages are a group of numbers that your program interprets as commands and data from Windows. The operating system calls a special function in your code that interprets the messages and dispatches them to the proper function.

When the user clicks on a button on a dialog box, for example, the click generates an *event*, and Windows sends your program a message about the click. A typical dialog box might generate any number of events from its controls. Even changing the selection in a list on a dialog box generates an event, and a message from Windows.

When you press and release a key on the keyboard, for example, Windows first notifies your program of the event through a message. Your program is free to respond to the event, or to ignore it altogether. C# programs contain default message handlers for many messages such as keyboard handling, so you do not need to worry about writing your own handlers.

Using Events

*As you will see shortly, you can add **message handler functions** to your code to respond to specific messages. You may see the term **event handler** in the documentation instead of message handler. In C#, events are more than just messages, although your program may treat them the same. Used in combination with **delegates,** your program can **publish** events. Normally in programming message handlers, if you handled the message in one class and you must pass the message on to another object, you had to write the code to relay the message.*

*C# carries that a step further with delegates and events that permit other objects to respond to the event. The object need only **subscribe** to the event, and you need not write the code to relay the message. You will learn about events and delegates in Lesson 29, "Using Delegates and Events."*

When you create a Windows application in C# using the Visual Studio, the wizard always adds a namespace to your source-code file using the project name. The namespace is not necessary, but it helps to isolate your code and to prevent you from duplicating names in the Visual C# library. There are more than 1,000 classes in the Visual C# library, and adding the namespace helps to qualify the names of your classes.

Visual Studio supports a large number of project types, and several languages, such as Visual Basic and Visual C++ in addition to Visual C#. To see a listing of the project types, start the Visual Studio. After the Visual Studio starts, select the File menu New item and choose Project. The Visual Studio, in turn, will dis-

play the New Project dialog box, as shown in Figure 26.1. (You also may click on Create New Project from the Start Page.)

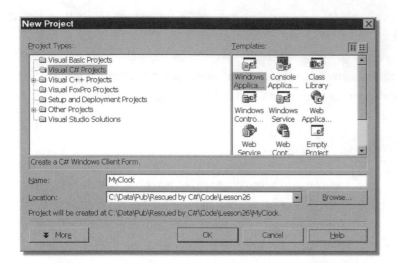

Figure 26.1. The New Project dialog box is where you will start building Windows applications in Visual Studio.

When you create a Windows application project, the Visual Studio always creates a form for you. A form is little more than a dialog box to which you will add the *controls* for your project. In Windows, a control is a special purpose window that you will use to display information to the user and to accept input from a user. The most common controls are the button and the scroll bar. You will learn more about controls and how to use them in Lesson 28, "Using Windows Controls."

In addition to the form, the Visual Studio creates the framework for the application type that you selected. Usually, this basic code is similar for most applications, and saves you a lot of time in preparing the basic source file.

Creating a Windows Application in Visual Studio

You know enough about the basics of C# programming to write applications. Now you must learn about the Visual Studio tools that will help you to write Windows applications. The best way to do that is by creating a Windows project in the Visual Studio.

The project you will build in this section will display a form on your screen that contains the current time. The time will update once every second, so you should see the seconds ticking away. If you are not running the Visual Studio, start it now. Then, perform the following steps:

❶ Select the File menu New option and choose Project. Visual Studio, in turn, will display the New Project dialog box.

❷ At the left of the New Project dialog box is a tree labeled Project Types. Within this list, you will select the programming language or project type you want to develop. For now, click your mouse on the item labeled Visual C# Projects.

❸ At the right of the dialog box is the Templates list, which contains the various projects that you can create using the language you selected in the previous step. Within the Template list, click your mouse on the Windows Application item.

❹ Using the Location field, select the directory where you want to store your project. You may use the Browse button to search for a directory.

❺ Within the Name field, enter the name of the project, which can be any name you want, but it should describe your project. By default, Visual Studio enters "WindowApplication" followed by a sequence number in this field, but that name is not very descriptive. In this case, type **MyClock**. Notice at the bottom of the dialog box that the Visual Studio adds the name of the project to the directory path for the project.

❻ Click your mouse on the OK button. Visual Studio will respond by creating your project.

Depending upon the speed and configuration of your computer, the Visual Studio may take some time to create the project. After the wizard has created the project, it will display a blank form as shown in Figure 26.2.

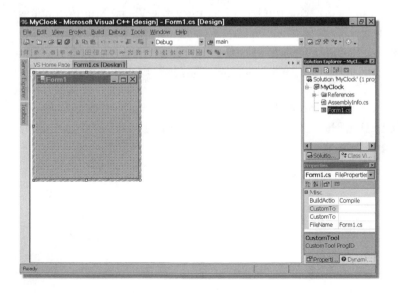

Figure 26.2. The initial display for the MyClock project shows a blank form along with several tool windows.

On the left side of the Visual Studio, you should see two tabs, one marked "Server Explorer" and the other marked "Toolbox." If the Toolbox tab is not visible, select the View menu and then select the Toolbox item to display the Toolbox. Make sure the Windows Forms item on the Toolbox is selected as shown in Figure 26.3.

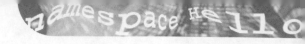

The items in the Windows Forms list are the various Windows controls that you may use on your form or in your program. For now, you are interested in only two items, the "Label" and the "Timer" items. The Timer item is near the middle of the list. Even on high-resolution displays, you probably will have to use the down arrow on the Toolbox to scroll the list until you see "Timer."

Double-click your mouse on the "Label" item. The Visual Studio, in turn, will add a small text field to your form. This is the control you will use to display the time.

If necessary, scroll until you find the "Timer" item and double-click your mouse on it. The timer is a "hidden" control, so the Visual Studio forms editor opens a small area at the bottom of the screen and displays the control in this area, as shown in Figure 26.4.

**Figure 26.4. Hidden controls
will appear in a separate area
just below the form.**

Next, expand the label box that you just added. To do this, move the mouse cursor over the lower right corner of the label box until the Visual Studio changes your cursor into the resize cursor, which appears as a double-ended arrow. Using the resize cursor, make the label box wide enough that it fills most of the form horizontally, and about one-fourth as deep as it is wide.

Next, click and hold the left mouse button somewhere in the middle of the label box and move it so that the label appears near the center point of the form, as shown in Figure 26.5.

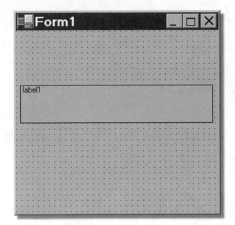

Figure 26.5. The MyClock project's form after resizing and moving the label box.

Before you leave the form, double-click your mouse on the timer icon in the box just below the form design area. The Visual Studio, in turn, will add a *message handler* function to your code where you may write the current time to the label box. Each time the timer sends your program a tick message, your program will enter this function and execute your code.

With the form prepared, you must turn your attention to the code. Press **F7** and the Visual Studio will open the source-code window for the form. You now should have at least two tabs at the top of the editing area and just below the toolbars. One should be labeled "Form1.cs [Design]," which displays the form, and the other "Form1.cs," which is the source-code window. You can move between these windows by clicking your mouse on one of the tabs.

Select the View menu, and then select the Class View item. The Visual Studio, in turn, will display the Class View tool window. Expand the tree by clicking on the "+" symbols until the tool window appears as shown in Figure 26.6.

Figure 26.6. The Class View tool window appears on the right side of your screen. The Solution Explorer also shares this tool window with Class View, and you may select between them by using the tabs at the bottom.

You now must add the function that will write the current time into the label box. Right-click your mouse on the Form1 item. The Visual Studio will display a pop-up menu. Within the menu, select Add, then Add Method. The Visual Studio will display the Add Method dialog box. Within the Add Method dialog box, type **SetTime** in the Method Name box. You do not need to worry about any other fields yet. Click your mouse on the Finish button that appears at the bottom of the Add Method dialog box.

Now, once again, turn your attention to the Form1.cs window that contains the source code. Locate the Form1 constructor function, *Form1()* and add code as shown in the following:

```
public Form1()
{
    //
    // Required for Windows Form Designer support
    //
    InitializeComponent();
    //
    // TODO: Add any constructor code after
    // InitializeComponent call
    //
    // The following line sets the text for the label box.
    // To display no text, simply use two quote marks.
    label1.Text = "";

// The following two statements set the text size and
// cause the time to appear centered in the label box
    label1.Font = new System.Drawing.Font
                            ("Microsoft Sans Serif", 24);
    label1.TextAlign = ContentAlignment.MiddleCenter;

// This.Text is the text that will appear in the form's
// title bar.
    this.Text = "My Clock";
// The following three lines set the timer to tick
// every second (1000 milliseconds), start the timer and
// write the initial time to the label box
    timer1.Interval = 1000;
    timer1.Start ();
    SetTime ();
}
```

Next, locate the *timer1_Tick()* function. This function is the message handler you added by double-clicking the timer icon earlier in this lesson. Within the function's code, add a call to *SetTime()*:

```
protected void timer1_Tick(object sender,System.EventArgs e)
{
    SetTime ();
}
```

Your code is nearly ready. The last step is to add the code to write the current time to the label box. You will place this code within the *SetTime()* method that you added to the code earlier:

```
SetTime ()
{
    string str = DateTime.Now.ToString();
    label1.Text = str.Substring (str.IndexOf (" ") + 1);
}
```

The *Now* property of the *DateTime* class extracts the current time. The string format of the time includes the date as well as the time in the form "7/29/2001 1:49:17 PM." You are interested only in the text following the first space, which is the time of day. You extract this information using the *SubString()* method of the *String* class. You then assign this substring to the *label1* box.

Your clock program is now ready for testing. You could have modified most of the code by using the Properties tool window. However, by modifying the code directly, you will learn where to look if you need to troubleshoot your program. You will learn how to use the Properties tool window in Lesson 28, "Using Windows Controls."

Compiling and Testing a Windows Program

Visual Basic programmers may recognize the clock window as a classic teaching tool for using forms in Visual Basic. Actually, the forms in Visual Basic and C# are the same, and you add the timer message handler and code in the same way.

Now that your code is ready, you must build, or compile, the program and run it in the Visual Studio debugger. To build the program, select the Build menu Build item. You also may use the Build toolbar that you learned about in Lesson 1.

When you start the build, the Output window will appear near the bottom of your screen. If all goes well, you should see the following message:

> **Build: 1 succeeded, 0 failed, 0 skipped**

To start your program in the debugger, select the Debug menu Start item. When the program starts, you will see your clock program similar to Figure 26.7.

Figure 26.7. When you run your clock program, it should display the current time of day, and update the time once a second.

If all does not go well—perhaps you made an error typing one of the lines of code—the Task List window will appear at the bottom of the Visual Studio window. Click on the *first* item in the Task List to locate and fix the error as you learned in Lesson 2, "Building, Running and Saving Your First C# Program." Always start with the first item when the Task List displays compile errors. A single error can generate several items in the Task List. Fix the first error, then recompile the program.

To experiment with the *Clock* project, try changing the size of the text in the label box, and try adding the date by extracting the first 10 characters from the time string. Add a second label box above the first label box to display the date.

WHAT YOU MUST KNOW

In this lesson, you learned how to create a Windows application using C#. You learned some of the ways that Windows programs differ from the command line programs that you have written, and that Windows uses the Graphic User Interface to display information in your window. You learned that Windows communicates with your program through a series of messages that your program may respond to or ignore. In Lesson 27, "Using Forms," you will learn how to add forms beyond the initial form that the Visual Studio creates for your application. Before you continue with Lesson 27, however, make sure you have learned the following key concepts:

- To create a Windows application using the Visual Studio, you use the New Project dialog box. From the dialog box, you may select the programming language and the type of project you want to create.

- The Visual Studio creates the framework for a Windows application, thus saving you a lot of time in preparing the basic code.

- For Windows applications using C#, the Visual Studio always creates the basic form, or dialog box, for your program.

- A control is a graphical object that you use to accept input from a user or to display information for the user. The toolbox contains several controls that you may add to your form.

- A form may contain many controls that communicate with your program through messages or events.

*i*n the last lesson, you created a Windows application that displayed a single form. The form simply displayed the current time in response to an event generated by a hidden timer. Your application is not limited to a single form, however, and a form is not limited to a single *control*. By using multiple forms, you can write some fairly complex applications using C#. In this lesson, you will develop an application that uses multiple forms and learn how to manipulate the forms from your program. By the time you finish this lesson, you will understand the following key concepts:

- A form is the C# equivalent of a dialog box. A C# Windows application always has at least one form.
- You may add additional forms to your C# application to display auxiliary information to the user or to accept input from the user.
- Forms may be *modal* or *modeless*. The user must dispatch a modal form before returning to another form. A modeless form may be used at the same time as other forms.
- You set the options for a form using the Properties tool window.
- The Toolbox tool window contains a list of the controls that you may add to a form.

Adding Forms to a Project

When you create a Windows application using C#, the Visual Studio creates a default form for you. You may create many powerful applications using just a single form. This is the equivalent of creating a dialog-based application.

More often than not, however, your program must use other forms to accept input from users or to display additional information the user may need while using the program. The form may be simply a message box informing the user of an event, or it may be another form containing Windows controls. In this lesson, you will create a project that will use both. If you have programmed in Visual Basic, adding and manipulating forms in C# is just as easy.

First, you must create a basic Windows application. Using the steps you learned in Lesson 26, "Writing a Windows Program," create a new C# Windows application named *MultiForm*. Before you do anything with the project, compile and run the program. You should see a blank form with nothing but the title bar containing the program icon on the left and the system buttons on the right.

Click your mouse on the "X" button to close the form and return to the Visual Studio. Using the Toolbar, double-click your mouse on the "Button" item. Visual Studio will add a button to the upper left corner of the form. You can move the button or resize it using the same steps you used in Lesson 26 to move and resize the label box.

Click your mouse on the button you just added to the form. Visual Studio will add an event handler, *button1_Click()*, to your source code. Add code to make it look like the following. *SecondForm* is the name you will give to the form that you will add to the project shortly:

```
protected void button1_Click (object sender,
                              System.EventArgs e)
{
    SecondForm second = new SecondForm ();
    second.ShowDialog (this);
}
```

Do not try to compile and run the program yet. The *SecondForm* class does not yet exist, and the C# compiler will issue an error if you compile the program.

If the Solution Explorer is not visible on the right side of the Visual Studio, select the View menu, then select Solution Explorer. Right-click your mouse on the MultiForm item, which should be the second item in the tree.

From the menu that pops up, select Add. Another menu will appear, so select Windows Form from this menu. The Visual Studio will display the Add New Item dialog box as shown in Figure 27.1.

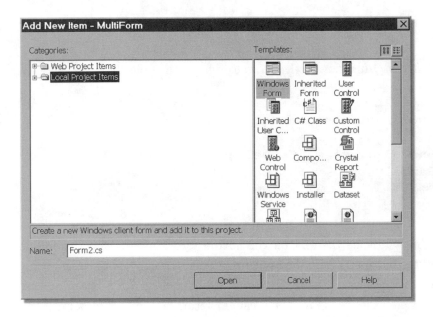

Figure 27.1. The Add New Item dialog box is an omnibus dialog box from which you may add controls, forms, datasets and other items to your project.

As you continue programming using the Visual Studio, you will become very familiar with the Add New Item dialog box. From this dialog box, you may add new forms, new controls, classes to access databases, or even just plain text files. On the left, select the Local Project Items.

On the right, the first two items in the Templates window are new forms. The first item, Windows Form, adds a basic form to your project. You use the second item, Inherited Form, to derive a new form from another. For now, you are interested in the first item, Windows Form. Select this item, then at the bottom of the dialog box, type **SecondForm** in the Name field.

Click your mouse on the Open button and the Visual Studio will add the new form to your project. You should notice that the source file for this form is *SecondForm.cs*. You had no choice when the Visual Studio created the first form for you, but when you add a form you can give them more descriptive names.

When you add a form using the Add New Item dialog box, Visual Studio will create the form and a new class for the form, and place the code in a separate source-code file.

Visual Studio will display the new form on a tab labeled "SecondForm.cs [Design]." Add one button to this form.

Press the F7 key to display the source code for *SecondForm.cs*. Locate the constructor and add the following call after the call to the *InitializeComponent()* function:

```
InitForm ();
```

Add the *InitForm()* function to the *SecondForm.cs* source file:

```
protected void InitForm()
{
    button1.Text = "Cancel";
    CancelButton = button1;
}
```

The first line will make the text on the button read "Cancel," and the second line makes the button perform the same function as the Cancel button on dialog boxes. In this application, the second form will be a *modal* form, meaning that when you display the second form you cannot return to the first form until you exit the second form. You could exit the second form by clicking the "X" button on the title bar, but the Cancel button will exit the second form gracefully.

Note: It is tempting to make the changes in the InitializeComponents() function. You should be aware that a comment in the source-code file tells you not to modify the InitializeComponent() method, but neither the comment nor the MSDN documentation explains why. If you change this function and later modify the properties of one of the controls using the Properties tool window, the Visual Studio will overwrite InitializeComponent() and your changes will be lost. That is why you add your own initializing function. In Visual C++, you normally would

initialize your controls in response to the WM_INITDIALOG message. That apparently does not work in Visual C# and I have not found a version of the OnInitDialog() function that you can override. There will be many times that you must initialize controls or other objects. You should not do this in InitializeComponents().

Compile and run your program. When you press the button on *Form1*, your program will display a second form as shown in Figure 27.2.

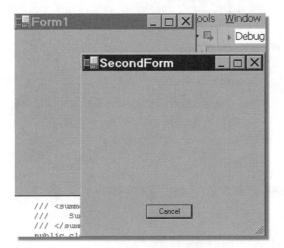

Figure 27.2. Clicking the button on the first form displays the second form. To close the second form, click on the Cancel button.

Close the second form by clicking on the Cancel button. If you press the button on Form1 again, you will create a new SecondForm. To exit the application, click on the "X" on Form1.

Understanding Modal vs. Modeless

*In C# you may create forms in two different modes, **modal** and **modeless**. When you create a modal form, the new form becomes the active form and you may not return to the first form until you dispatch, or close, the first form. A modeless form, however, becomes the active form, but you may return to the first form before you dispatch the second form. If you try to return to the first form, Windows will just beep the speaker on your PC.*

*Modal forms are the most common, and you use the mode when you need to have the user enter some information that the first form needs. You create a modal form with the **ShowDialog()** function. Modeless forms are like auxiliary forms. The user may use both forms at the same time. When you create a modeless form, you may return to the first form by clicking your mouse on it, and the second form will remain visible. You create a modeless form using the **Show()** function.*

*In the project in this section, change the **button1_Click()** to call **second.Show()** rather than **second.ShowDialog()**. Compile your program and run it again. Click your mouse on the button to create the second form. You now can return to the first form while leaving the second form visible. Notice, however, that the Cancel button no longer will close the second form. To close the second form, you need to click on the "X" button on the title bar. If you want the Cancel button to work, you must add an event handler to the source code. To do this, return to the **SecondForm.cs** design page and double-click your mouse on the button. This will add the event handler to your code. Make the event handler read as shown below:*

```
protected void button1_Click (object sender,
                                 System.EventArgs e)
{
    Close ();
}
```

Recompile and run your program again. The Cancel button once again will close the second form.

Setting Form Options

The default form C# creates may not always be to your liking, or meet the needs of your program. It may not appear as you would like, for example. Forms have *properties* that control their appearance and operation, and you may modify these properties from within the Visual Studio Form Designer. You may modify the form properties by using the Properties tool window.

Using the same steps as before, create a new C# Windows application named "Properties." When the Visual Studio displays the form in the Form Designer, right-click your mouse on the form and select Properties from the menu that appears. This will display the Properties tool window in the lower right corner of your screen, as shown in Figure 27.3.

Figure 27.3. The Properties Tool window is where you will set options for forms and controls.

Using the scroll bar on the right side of the Properties tool window, scroll until you find the "Text" entry and click your mouse on it. The default value for the Text property is "Form1," which is the text that will appear in the title bar for the form. Click your mouse on the text and change it to read "My Form."

Next, scroll until you find the BackColor property and click on it. This is the background color for the form, and it defaults to "Control," which is a light gray. Notice that a small down arrow appears on the right side of the line that displays the property. This arrow indicates a menu from which you may select other values. Click your mouse on the down arrow and select another color such as "Window" for the form.

Now scroll until you find the Font property and click on it. The Font property is the font the Form Designer will use to display text in the controls that you add to the form. The default is Microsoft Sans Serif. Changing the Font property will not change the font for the title bar. Notice here that a small button with an ellipsis appears on the right side of the line. The ellipsis indicates that clicking the button will display a dialog box from which you may select or enter a new property. Click your mouse on the button showing the ellipsis and the Form Designer will display a Font dialog box. Select another font such as Times New Roman. Return to the form and add a button to the form. You will see that the text on the button is in the new font that you selected. Add a label, and the text will appear in the new font as well.

Compile and run the project. When the form displays, you will see the form using the properties that you set.

The Properties tool window has a small help window attached to it that gives a brief explanation of each property. If the Help window is not visible, move the mouse to the bottom of the tool window until the cursor changes into two vertical lines with handles. Click your mouse and hold the left mouse button and move the cursor upward. The help window will appear, and each time you click on a property, a short description of the property will appear.

Using Inherited Forms

In the previous topic in this lesson, you learned that the Add New Item dialog box contains two entries to add new forms to your project, a Windows Form item and an Inherited Form item. Now that you have learned how to customize the properties of a form, you may use the Inherited Form item to create new forms using your custom properties.

If you did not make the changes to the **Form1** properties in this topic, go ahead and do it now. Using the steps you learned earlier in this lesson, add a second form to the project, except this time select the Inherited Form item instead of the Window Form item, then click on the Open button.

Instead of adding the form to your project immediately, the Form Designer will display an Inheritance Picker dialog box as shown in Figure 27.4. The list in this dialog box displays existing forms in your project, and you may create the new form using the properties of one of these forms.

Figure 27.4. The Inheritance Picker dialog box lists the names of existing forms in your project.

Select the Form1 item and click your mouse on the OK button. The Form Designer will create the new form using the properties of Form1. Using the steps in the previous topic, add an event handler for the button on Form1 to display the second form when you click on the button.

Compile and run your program again. Click your mouse on the button to display the second form. The new form should be identical to Form1. The new form also will inherit the methods and event handlers that you added to **Form1.cs**. You may modify the properties and override the methods in the new form without affecting the original form.

The Browse button on the Inheritance Picker allows you to open another project and inherit the properties from a form in that project. Thus, when you design a special form and set its properties, you do not have to design it again for every project.

Displaying and Hiding a Form

Forms may exist in your project and not always be displayed. In this lesson, you created a project with more than one form, and you created that form in the event handler for a button on the first form. Each time you click the button, you create a new instance of the second form's class. Using this method, you re-create all the controls on the form at the same time, and your program will not retain all the changes to the form.

One problem with using this method comes when you create the second as a modeless form—you may return to the first form and create another instance of the second form. Thus, you have two copies of the second form displayed when you only want one.

By creating a form before you need the form, you can hide and display the form at any time. If you simply hide the form, any changes to the form will appear when you display the form again.

Create a new project named "Display." The project should be identical to the *MultiForm* project from earlier in this section. You may use the *MultiForm* project if you want. You will modify the project later to hide and redisplay the form without re-creating it. Be sure to add a button to Form1 and an event handler to create and display the form when you press the button. Use the *ShowDialog()* method to display the second form as a modal form.

On the second form, add a text box control. With the second form displayed in the Form Designer, double-click your mouse on the TextBox item on the Toolbox to add the text box. Move the text box to a convenient location on the form and resize it if you want. Do not change the default text "textbox1."

Compile and run your program. Click your mouse on the button to create the second form. Type some changes in the text box, then click the button to close the form. When you return to the first form, click the button again to redisplay the second form. The changes you made in the text box are lost, and the default text appears again.

Exit the program and return to the *Form1.cs* source-code file. Change the event handler as follows:

```
protected void button1_Click (object sender,
                                  System.EventArgs e)
{
   second.Show ();
}
```

Now add a field to *Form1.cs*. Simply type the following line in the *Form1* class definition, but *outside* any function or method definition:

```
SecondForm second = new SecondForm();
```

Select the *SecondForm* in the Form Designer. Double-click on the button to add an event handler for the button. Make the event handler in *SecondForm.cs* read as follows:

```
protected void button1_Click (object sender,
                                  System.EventArgs e)
{
    Hide();
}
```

Now recompile and run your program again. Click your mouse on the button to display the second form, then make some changes to the text in the text box. Click your mouse on the Cancel button (or button1 if you did not change the button text). The second form will disappear and you will return to the first form. Click your mouse on the button on the first form again and the second form will reappear, and the changes you made to the text box will be there.

In this project, you display the second form as a modeless form. You could use a modal form by calling *second.ShowDialog()* as well. However, if you return to the first form and click the button again, instead of creating and displaying another instance of the second form, nothing happens.

Your program creates the second form when it first runs, but it does not display the form. The event handler for the button on *Form1* displays the form, and the event handler for the button on *SecondForm* hides the form. At the same time, your program will retain any changes to the form, and you prevent accidentally creating a second instance of the form.

Adding Controls to a Form

In Lesson 28, "Using Windows Controls," you will learn about controls and how to create and initialize them. Before you leave this lesson, however, you should explore the various methods of adding controls to a form.

Until now, you have simply double-clicked your mouse on a control item on the Toolbox to add a control. This places the control in the upper left corner of the form, and you then must move it to another location.

You also may drag and drop a control onto a form. Using this method, the Form Designer will create the control at the location where you drop it. To drag and drop a control, simply click your mouse on the item in the Toolbox (you do not have to hold the button down as you do in most drag and drop operations). Move the mouse cursor over the form and click and hold the mouse where you want the upper left corner of the control to appear. Draw the control to the size you want and release the mouse button. You may resize the control and make fine adjustments to the location using the cursor control keys.

To use drag and drop, you probably will want to turn off the auto-hide feature of the Toolbox. If auto-hide is on, the Toolbox will appear and cover a part of your form, sometimes making it impossible to drop the control into the proper position.

To turn off auto-hide, right-click on the Toolbox title bar. From the menu that pops up, select the the Auto Hide item to uncheck it. To restore auto-hide, perform the same step to check the auto-hide.

You also may create controls dynamically, writing code yourself to create and initialize a control. You will have to set the size and position on your own. The following snippet adds a button to the center of a form:

```
Button MyButton = new Button ();
private void InitMyButton()
{
    // Set the text of button1 to "Cancel"
    MyButton.Text = "Cancel";
    // Set the text color to the default
    MyButton.ForeColor =
                    System.Drawing.SystemColors.ControlText;
    // Set the position of the button to the center of
    // the form
    MyButton.Location = new Point
                    (this.ClientRectangle.Width / 2 -
                     MyButton.ClientRectangle.Width / 2,
                     this.ClientRectangle.Height / 2 -
                     MyButton.ClientRectangle.Height / 2);
    // Set the text to "Cancel"
    MyButton.Text = "Cancel";
    // Add the button to the form
    Controls.Add(MyButton);
    // Set the button as the Cancel button
    CancelButton = MyButton;
}
```

You then would have to place a call to *InitMyButton()* in the constructor function, or call *InitMyButton()* function when you want to create the button.

Rolling your own controls can be tedious and the arithmetic sometimes can get involved, but there are times when you do not know in advance what controls you will need on a form, or the size of those controls.

WHAT YOU MUST KNOW

Every C# Windows application begins with at least one form, and you may add more forms as you need them. In this lesson, you learned how to create additional forms for your C# Windows application. You also learned how to display forms as modal and modeless. In addition, you learned how to add controls to your form using the Toolbox tool window, and how to dynamically add controls by creating them in your code. In Lesson 28, "Using Windows Controls," you will learn how to use several of the common controls in your C# program. Before you continue with Lesson 28, however, make sure you have learned the following key concepts:

- The Visual Studio creates your program's first form when you create a C# Windows application.

- Your program may need more than one form. You can add forms by using the Add New Item dialog box.

- You can customize the properties of a form by using the Properties tool window.

- Once you create a custom form, you may create forms with the same properties by using the Inherited Form item on the Add New Item dialog box.

- Using modeless forms, the user may switch between forms without first having to close one form.

USING WINDOWS CONTROLS

Y ou have learned how the Visual Studio creates a form when you create a C# Windows application, and how to add and display additional forms in your project. A form by itself, however, is not of much use. You must provide some means of displaying information to the user, or accept input from the user. To do this, you must add controls to the form. In Lesson 28, "Using Forms," you saw that even to test a form you needed basic controls such as a button and a text box. In this lesson, you will learn about many of the Windows controls and how to use them on a form. By the time you finish this lesson, you will understand the following key concepts:

• A form is a container for Windows controls. You use the controls on a form to accept input from a user or to display information.

• Windows includes a library of *common controls* that may be used by any programming language to write Windows programs.

• You may modify the basic properties of a control by using the Properties tool window.

• Windows communicates with controls through messages. The controls communicate with your program through messages as well. Your program may respond to many of these messages through *event handler* functions.

• Event handler functions are *delegates* that notify your program of important events occurring in controls.

Understanding the Purpose of Controls

A form is a dialog box, but it is not much more than that. A dialog box is a container for Windows controls, which you use to communicate back and forth with the person using your program. You can create a form with only basic controls such as a title bar and a system menu, but from the user's point of view, this type of form does not help to achieve the task at hand.

To make a form useful, you must add controls to the form. A control is a graphic object that displays information in some form such as text or a list, or accepts information from a user, such as a button click or input from the keyboard.

In programming a Windows application, you usually create a window. A form is a window with a special purpose. Virtually everything you see on or in the window is a control. The title bar is a control, the menu is a control, system buttons to maximize, minimize and close a window are controls.

Windows provides a set of *common controls* that any program running on a Windows system may access and use. Some of these controls have been available for many years and other, more advanced controls were introduced when Microsoft released Windows 95. Periodically, Microsoft updates the common control library and includes new controls.

Controls relieve the programmer of many of the tasks of writing a Windows program. The controls include much of the basic code a programmer needs to use them. If Windows did not provide the common controls, programmers would invent a similar library, if for nothing else than to ease the programming tasks. The text box control, for example, contains all the programming to accept keyboard input, insert characters into a string, delete characters and display the resulting text. The programmer may want to add more capability to the control, but the basic code already is in place.

Controls are Multilingual

Whether you write Windows applications in C#, Visual Basic, Visual C++ or any of a number of other programming languages, you will need to know about the Windows common controls library. Even World Wide Web languages such as Java use the common controls library.

*The common controls library is a **dynamic link library** (a "dll") that your program loads into memory when it must use one of the common controls. It has standardized entry points that may be called from any language.*

Microsoft updates the common controls library periodically, adding new controls and updating and fixing bugs in existing controls.

Examining Common Controls

Microsoft has provided programmers with a set of basic controls since it first introduced Windows. Over the years, the controls have been refined and their number has grown. These *common controls* present the user with a consistent interface. Although the programmer may customize the common controls to some extent, the common controls function the same in every program that uses them. The user does not need to learn new keyboard commands to use a control in every program.

Visual C# implements the common controls, but not always under the same name many programmers may recognize. Originally, Windows contained six basic controls, as summarized in Table 28.1 along with the MFC class that supports them.

Control	C# Name	Description
Static Text	Label	A text control for labeling other controls.
Edit Box	TextBox	Boxes for entering text.
Button	Button	Buttons such as OK or Cancel.
Group Box	GroupBox	A button control used to group other controls visually.
List Box	ListBox	A list of strings.
Combo Box	Combobox	Edit box and list box combination.
Vertical Scroll Bar	VScrollBar	Vertical scroll bar used inside a dialog box.
Horizontal Scroll Bar	HScrollBar	Horizontal scroll bar used inside a dialog box.

Table 28.1. The basic Windows common controls and their C# names.

Another common control, the Radio Button control, is a special type of the button control. When used in a group, the radio button control allows the user to select only one button in the group.

To explore the basic common controls, create a new C# Windows application named "Controls." The controls you will need to add to it are shown in Figure 28.1. First, you will have to drag the OK and Cancel buttons from the right side of the dialog box to the bottom.

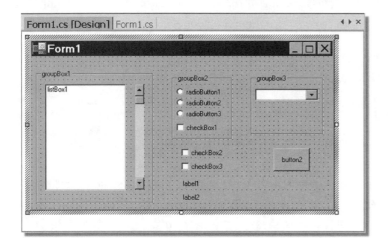

Figure 28.1. The Controls form demonstrates how you might use some of the basic common controls.

Add controls to the dialog box in the following order. To draw a control, click your mouse on the item in the Toolbox. Then move the mouse to the form and left-click and hold the mouse button where you want the upper left corner to appear. Drag the mouse to the where you want the lower right corner of the control to appear. Release the mouse button. Refer to Figure 28.1 for the location of the various controls on the form.

❶ Draw the three group boxes first. You can resize the group boxes later if necessary, but you should draw them first. If you draw these after you draw the controls within them, the group box may cover the other controls. If this happens, select the group box, then select the Format menu, then Order and Send to Back. The controls underneath should appear.

❷ Draw the list box inside the first group box.

❸ Draw a vertical scroll bar inside the first group box next to the list box.

❹ Draw three radio button controls in Positions 3, 4 and 5. Draw these one after another so the Visual Studio will group them so they work together. In Visual C++, you would set the Group property on the first radio button, then uncheck Group on the other controls. There is no Group property in Visual C#, however. These buttons will be *radioButton1*, *radioButton2* and *radioButton3*.

❺ Draw a check box just below the three radio buttons. This will be *checkBox1*.

❻ Draw two check boxes below the group box that contains the radio buttons. These will be *checkBox2* and *checkBox3*.

❼ Draw the combo box inside the third group box. Draw a button below the group box. The button will be the application's Finished button. Finally draw two label fields across the bottom of the form as shown in Figure 28.1.

Double-click your mouse on each of the following items to add event handlers for the controls: the list box, the check box just below the radio buttons, the two check boxes below the radio button group box, the combo box and the button. Do *not* double-click your mouse on the radio buttons. You must add event handlers for these buttons manually.

You should pay attention to the event handlers for the radio buttons in the following code. The MSDN help file does not give much information about using radio buttons in a C# program, and adding the event handlers does not create the code to enable the events. You must add *delegates* to your initialization code to enable the events.

You must add *InitForm()* and *radioButtons_CheckedChanged)* methods to the *Form1* class as shown in the following code. At the same time, add code to the event handlers as shown (the code generated by Visual Studio is not shown in the following listing):

```
// Checking one check box disables the other. Unchecking
// the box enables the other.
protected void checkBox2_CheckedChanged (object sender,
                                          System.EventArgs e)
{
    if (checkBox2.Checked == true)
        checkBox3.Enabled = false;
    else
        checkBox3.Enabled = true;
}
protected void checkBox3_CheckedChanged (object sender,
                                          System.EventArgs e)
{
    if (checkBox3.Checked == true)
        checkBox2.Enabled = false;
```

```csharp
    else
        checkBox2.Enabled = true;
}

protected void button1_Click (object sender,
                               System.EventArgs e)
{
    Close ();
}

protected void radioButtons_CheckedChanged (object sender,
                                             System.EventArgs e)
{
    if ((sender == radioButton1) &&
                    (radioButton1.Checked == true))
    {
        checkBox1.Enabled = true;
        checkBox1.Checked = false;
    }
    else if ((sender == radioButton2) &&
                    (radioButton2.Checked == true))
    {
        checkBox1.Enabled = true;
        checkBox1.Checked = true;
    }
    else if ((sender == radioButton3) &&
                    (radioButton3.Checked == true))
    {
        checkBox1.Checked = false;
        checkBox1.Enabled = false;
    }
}

protected void comboBox1_SelectedIndexChanged(object sender,
                                              System.EventArgs e)
{
    label1.Text = "Combo Box selection is now " +
                        comboBox1.SelectedItem.ToString();
}

protected void listBox1_SelectedIndexChanged (object sender,
                                              System.EventArgs e)
{
```

```
            label2.Text = "List Box selection is now " +
                              listBox1.SelectedItem.ToString();
}
//
// InitForm() initializes the controls
protected void InitForm ()
{
    string [] str = new string [50]
    {
        "Alabama", "Alaska", "Arizona", "Arkansas",
        "California", "Colorado", "Connecticut",
        "Delaware", "Florida", "Georgia",
        "Hawaii", "Idaho", "Illinois",
        "Indiana", "Iowa", "Kansas", "Kentucky",
        "Louisiana", "Maine", "Maryland",
        "Massachusetts", "Michigan", "Minnesota",
        "Mississippi", "Missouri", "Montana",
        "Nebraska", "Nevada", "New Hampshire",
        "New Jersey", "New Mexico", "New York",
        "North Carolina", "North Dakota", "Ohio",
        "Oklahoma", "Oregon", "Pennsylvania",
        "Rhode Island", "South Carolina", "South Dakota",
        "Tennessee", "Texas", "Utah", "Vermont",
        "Virginia", "Washington", "West Virginia",
        "Wisconsin", "Wyoming"
    };
// Label the form
    Text = "Controls";
// Set the text for the group boxes
    groupBox1.Text = "List Box and Scroll Bar";
    groupBox2.Text = "Buttons";
    groupBox3.Text = "Combo Box";
// Add the state names to the list box and combo box
    foreach (string s in str)
    {
        comboBox1.Items.Add (s);
        listBox1.Items.Add (s);
    }
// Make the combo box show the first item in the list
    comboBox1.SelectedIndex = 0;
// Check the first radio button, then set them for
// auto check so they will show the check mark.
    radioButton1.Checked = true;
```

```
    radioButton1.AutoCheck = true;
    radioButton2.AutoCheck = true;
    radioButton3.AutoCheck = true;
// When you add the event handler for a radio button
// check, Visual Studio does not add the delegate.
// Create the delegates using the three lines below.
    radioButton1.CheckedChanged += new System.EventHandler
                        (this.radioButtons_CheckedChanged);
    radioButton2.CheckedChanged += new System.EventHandler
                        (this.radioButtons_CheckedChanged);
    radioButton3.CheckedChanged += new System.EventHandler
                        (this.radioButtons_CheckedChanged);
// Label the radio buttons
    radioButton1.Text = "Button up";
    radioButton2.Text = "Button down";
    radioButton3.Text = "Disable";
// button1 will be the exit button. Label it "Finished"
    button1.Text = "Finished";
// Make the check box look like a button
    checkBox1.Appearance =
                        System.Windows.Forms.Appearance.Button;
}
```

Selecting one of the radio button controls should change the appearance of the button control just below the radio buttons. The scroll bar does nothing except move up and down. Selecting an item in the list box control or combo box control should display a message box in the labels on the lower right of the form telling you the current selection.

Eventually, the basic controls were not enough. With the advent of multitasking versions of Windows, a number of controls were added to the common controls library. For example, when you open a "folder" in Windows 95 and above, the operating system displays the contents of the folder using the new List control (do not confuse the List control with the basic List Box control).

Table 28.2 summarizes many of the more advanced common controls. This is not an exhaustive list. Microsoft has added a number of controls to the library recently, and others just do not appear on the toolbar.

Control	C# Name	Description
Date/Time Picker	DateTimePicker	Interface to display and enter date and time information.
Image List	ImageList	Is used to manage large sets of icons or bitmaps. (This is not a true control, but it supports lists used by other controls).
List	ListView	Displays a list of text with icons.
Progress	ProgressBar	A bar that indicates the progress of an operation.
Rich Edit	RichTextBox	Edit control that allows multiple character, paragraph and color formatting. Often used as the underlying control for a document window.
Slider	TrackBar	Similar to a sliding control used as volume control on audio equipment.
Spin Button	NumericUpDown	A pair of arrow buttons to increment or decrement a value (C# combines this control with an edit box).
Status Bar	StatusBar	A bar to display information such as the state of the insert or NumLock keys, or to write status or help messages.
Tab	TabControl	Is used in property sheets. Similar to notebook tabs.
Toolbar	ToolBar	Contains buttons to generate command messages.
ToolTip	ToolTip	A small pop-up window that describes the use of a button or tool.
Tree	TreeView	Displays a hierarchical list.

Table 28.2. The Windows advanced common controls.

You should be aware of these controls so you may recognize when they are used in other programs. For example, in the Visual Studio the Solution Explorer and Class View use the tree control to list the objects in your program.

Setting Control Styles

Up to now, you have set the styles—the properties—of the controls on your form in code. For example, when you set the text for a group box in the program in the preceding section, you wrote something like the following:

```
groupBox1.Text = "List Box and Scroll Bar";
```

Sometimes this means repeating code that the Visual Studio placed in the *InitializeComponent()* method of your form class. There are times when you must set a property in your initialization code, but for many properties you may use the Properties tool window. By default, the Properties tool window appears in the lower right corner of the Visual Studio. This is the same properties window that you used to set properties for a form in Lesson 27, "Using Forms."

Open a C# Windows application in the Visual Studio and select a form in the Form Designer. The Properties tool window should appear. If it does not, right-click your mouse anywhere on the form and select Properties from the menu that pops up. You also may select the View menu, then the Properties Window item.

Setting properties for individual controls can be a daunting task until you become familiar with the controls and the Properties tool window. Sometimes it is just easier and quicker to add the controls in an initialization function.

To use the Properties tool window with a control, simply select the control on the form. The tool window will change to reflect the properties available for that particular control. Some controls such as the Label and the GroupBox have only a few properties. Some of the more advanced controls such as ListView have many properties.

After you select a control, scroll through the properties to find the one you want to set. Some of the properties open menus or dialog boxes similar to those you encountered in Lesson 27.

The Properties tool window has its own toolbar located at the top of the tool window and shown in Figure 28.2. From left, the five buttons on the toolbar select a listing by category, an alphabetized listing, a listing of just the properties, a listing of just the events for an object and a button to display the property pages for an object. The last button generally is not available for controls and is disabled.

Figure 28.2. The Properties toolbar speeds selection of the various categories of properties.

If you know the name of a property, an alphabetic listing probably would be faster. Otherwise, a categorical list is more convenient. If you need to make the

Properties tool window larger, you can shrink the Solution Explorer or the Class View windows by grabbing the divider between them and moving it up, or you can simply hide the Solution Explorer and Class View windows.

Using Messages From Controls

Windows communicates with your program and with the controls on a form through messages. A message often is an *event*, but there are many messages that do not necessarily correspond to events. The messages that you are most interested in, however, are events.

For example, you might want to know when a user clicks his or her mouse on a button such as the Accept (or OK) button so that your program can check the contents of the various controls on the form. However, many messages sent in the form of notification messages are housekeeping messages, and do not have entries in the Events listing of the Properties tool window.

The Visual Studio will add the most common or most useful event handler to your code when you double-click your mouse on a control in the Form Designer. For a button, this will be a click event, and for a check box or radio button, it will be an event created when the check mark changed. For other controls such as the timer control, the click and check-changed events have no meaning, so Visual Studio will add a tick event handler to your code when you double-click your mouse on a timer control.

For other events, you must use the Events listing of the Properties tool window. Select the control on the form in the Form Designer, then select the Events listing in the Properties window (the icon showing a lightning bolt). From the Events listing there are two ways to add event handlers. In each case, you click your mouse on the area to the right of the event name.

❶ Select an existing event handler. When you click your mouse on the blank area next to the event name, a down arrow button will appear. Click your mouse on this button and a menu containing the names of other event handlers in your form class will appear.

❷ Create your own event handler name. Single-click your mouse on the blank area next to the event name and enter the name of the function that you want to be the event handler. Visual Studio will create the function and make it the event handler.

In addition to adding the event handler function to your code, the Visual Studio will add a statement to the *InitializeComponent()* function similar to the following:

```
listBox1.SelectedIndexChanged += new System.EventHandler
            (this.listBox1_SelectedIndexChanged);
```

This is an example of a *delegate*. Using delegates, you may specify more than one event handler for a single event, including event handlers in other classes. You will learn more about how to create and use delegates in Lesson 29, "Using Delegates."

WHAT YOU MUST KNOW

In this lesson, you learned about many of the Windows common controls and how to add them to a form in your Windows application. You learned how to modify the properties, both by adding your own code in an initializing function, and by selecting them in the Properties tool window and entering new values. You learned how to add event handlers to respond to messages from your controls about important events. In Lesson 29, "Using Delegates," you will learn more about writing event handlers and using delegates, including how to write multiple handler functions for a single event. Before you continue with Lesson 29, however, make sure you have learned the following key concepts:

- To make a form usable, you must add controls to it. You use controls to communicate with the program's user.

- The Windows common controls library may be used by a variety of programming languages other than C#.

- The Properties tool window contains entries to modify the properties of your forms and controls.

- You also may add event handler functions by using the Properties tool window.

- You may create your own event handler functions and add them through the Properties tool window.

USING DELEGATES

ontrols may generate events, as you learned in Lesson 28, "Using Windows Controls." You may handle these events in your C# program by defining an event handler method. Just defining the event handler is not enough, however. The Common Language Runtime does not know the name of the method in which it will handle the event. You must tell the Common Language Runtime the method name by creating a *delegate* which you pass to the the Common Language Runtime. In this lesson, you will learn how to define your own delegates, both in command-line programs and in a C# Windows application. By the time you finish this lesson, you will understand the following key concepts:

• A *delegate* contains a reference to a function or method. A delegate defines a method but does not implement the method.

• Delegates are similar to function pointers, but are type-safe. They may not call methods or access fields they are not authorized to access.

• Using delegates, you may assign the function that your program will call.

• Delegates may be single-cast or multi-cast. To use a multi-cast delegate, the referenced method must have a return type of *void*.

• Event handlers are delegates. Event handlers always are implemented as multi-cast delegates.

Understanding Delegates

There are times when the runtime code—the Common Language Runtime in the .NET environment—must call a function in your program. This happens when a control generates an *event* and your code must execute some statements to handle the event. For example, the person using your program presses a key on the keyboard. Your code, in turn, must process that key, determining whether it is a character that must be displayed or whether it is a function key that requires some action by your program.

At other times, the Common Language Runtime may need to call your program to get some information. The list control, for example, may list a number of items in one or more columns. You can make the list control sort the entries in the list by using the values in any of the columns. However, the Common Language Runtime does not know about the entries in the list before you write your program, and thus cannot know how to sort the items. In C++, you would write a *callback* function, and the runtime code would call this function to get the relationship between two items, and thus find out how to sort the list.

Normally, you would pass the runtime code the address of a function that you want to handle the event or the callback. Essentially, this is a *pointer* to the function, and the use of function pointers is common in C++. However, C# does not let you use pointers, except in *unsafe* code. C# uses *references* in places where you would need to use a pointer in another programming language, and you create a reference object using the *new* operator. The *new* operator creates an instance of the object in the heap, and returns a pointer to the object. C# then uses this pointer as a reference to the memory address in the heap where the object lives.

It would be a difficult task, however, to write a programming language that created functions dynamically on the heap. To get around this problem, C# uses the *delegate* concept to replace function pointers. Delegates are not new with C#, and Visual C++ programmers may be familiar with delegates.

A delegate essentially defines a function or method without implementing the function. To use a delegate, you must replace that definition with a method of your own and then tell the runtime code where in your program to find the method. Then the runtime code knows what statements to execute at the proper time.

Thus, the closest equivalent of a delegate in C and C++ is a function pointer, and in fact a delegate is the object-oriented equivalent of a function pointer. Delegates actually are an improvement on the function pointers, however. Function pointers may reference only *static* functions, and methods such as call-back functions thus do not have access to non-static members of a class. A delegate may reference either *static* or non-*static* member functions. When you use a non-*static* method, the delegate stores a reference to the class object as well as the function.

In addition, delegates in C# are *type-safe*, meaning they may access only code that they are authorized to access. They cannot, for example, access another object's private fields or methods, even in the class in which you declare the delegate.

Using Delegates

A delegate is the C# equivalent of a function reference. When you use a delegate, it tells the runtime code where in your code to find the function that it must execute. The best way to learn about delegates is to examine a program that uses a delegate. The following program, *SinglCst.cs*, is the simplest implementation of a delegate. The program declares a delegate, and then assigns a function reference to the delegate first using a *static* method and then using a non-*static* method. In between, it calls the *ExecuteDelegate()* method, which executes different functions:

```
namespace nsDelegates
{
    using System;
    class clsMain
    {
        public delegate void TextHandler (string str);
        public TextHandler ShowText = null;

        public void ExecuteDelegate ()
        {
```

```
            if (ShowText != null)
                ShowText ("Delegate called");
        }
        public static void Main ()
        {
            clsMain main = new clsMain();
            main.ShowText = new TextHandler (Static);
            main.ExecuteDelegate();
            main.ShowText = new TextHandler (main.Instance);
            main.ExecuteDelegate();
        }

        private static void Static(string str)
        {
            Console.WriteLine ("Static() Method: " + str);
        }
        private void Instance(string str)
        {
            Console.WriteLine ("Instance() Method: " + str);
        }
    }
}
```

The first line in *clsMain* declares the delegate and sets up a template—a pattern—for the type of method that may be used with the delegate. It effectively declares a function but does not provide a body of code for that function. In this case, you are telling the C# compiler that statements that use this delegate must pass a *string* argument.

The second line declares a field—a variable—in which to hold the function reference. The data type is *TextHandler*, the delegate type that you declared in the previous line. The function reference that you assign to this field must conform to the pattern you declared in the delegate. For example, the *TextHandler* pattern requires a *string* parameter, so you could not assign a function reference to *ShowText* that requires an *int* parameter. You set the value of *ShowText* to *null* to indicate that it does not initially contain a function reference. Technically, you do not have to assign *null* as an initial value because the C# compiler automatically initializes instance fields to 0 or *null*. Its presence in the code is just to emphasize that it is *null* and that you may assign it a *null* value. This is equivalent to assigning *Nothing* to an object in Visual Basic.

Next, you declare and define the method that you will use to access the delegate, *ExecuteDelegate()*. In this method, you first test whether *ShowText* is *null*. If it is, you do nothing. If you attempt to call this method before you assign a reference to *ShowText*, your program will throw a *NullReferenceException*. You could define *ExecuteDelegate()* so that it accepts parameters that you may pass on to the delegate, or to return a value. You will do this shortly.

These three steps set up your delegate, so now you must use it. In the *Main()* function, you first create an instance of *clsMain*, the class that contains the delegate. You should note that you could have declared the members *static* in the preceding steps and used them without declaring an instance of the class, as you will do in the next example.

In the next line, you create an instance of the delegate. Notice that you must use the *new* operator to do this. The *delegate* keyword actually is an alias for the C# runtime's *Delegate* (or the *MulticastDelegate*) class, and so the *TextHandler* variable actually is a reference-type variable. You must pass the constructor for *TextHandler* a reference to a method that matches the pattern you specified when you declared the delegate. You cannot pass the reference method any parameters in this declaration, and you write the method's name without a parameter list. You simply write *Static* instead of *Static("string")*.

Now that you have created the delegate and assigned the variable a reference to a method, you may call the *ExecuteDelegate()* method, which in turn will call whatever method you have assigned to *ShowText*.

The next two lines repeat the previous two lines, but assign a non-*static* method to *ShowText*. Notice that you must include the instance name, *main*, as part of the non-*static* method's name.

When you run this program, you should see the following two lines of output, showing that in each call to *ExecuteDelegate()* your program called a different function:

```
Static() Method: Delegate called
Instance() Method: Delegate called
```

The runtime *Delegate* class maintains a list of methods that have *subscribed* to the delegate. You may assign more than one method to the delegate by using the += operator in the assignment. Try changing the code in *Main()* to the following:

```
public static void Main ()
{
    clsMain main = new clsMain();
    main.ShowText += new TextHandler (Static);
    main.ShowText += new TextHandler (main.Instance);
    main.ExecuteDelegate();
}
```

Compile and run *SinglCst.cs* again and you will see the same output, even though you call *ExecuteDelegate()* only once. To *unsubscribe* from a delegate, use the -= operator instead of the += operator. For example, if you want to unsubscribe the *Static()* method, you would use the following line:

```
main.ShowText -= new TextHandler (Static);
```

This program is a trivial example of a delegate. There are times that you may want to use delegates in this way, but you do not take advantage of one of the more important properties of delegates, type safety. By declaring the delegate in the same class in which you use it, the *ExecuteDelegate()* function could call the methods directly, even through you declared the methods *private*. For example, you could add the following statement to *ExecuteDelegate()* and it still would execute:

```
Instance("This is wrong!");
```

Normally, you will want to write a class to hold the delegates, which would prevent such nastiness.

Writing a Class for Delegates

In the last topic, you wrote the delegate as a member of the *clsMain* class. Without declaring an instance of *clsMain* or changing the members that set up the delegate to *static*, you cannot use delegates outside *clsMain*. By placing your delegates in a separate class, you may use them whenever you need them simply by declaring an instance of the class.

The following program, *DlgClass.cs*, shows how you would set up a delegate class, then declare an instance of the class to execute the delegate:

```
namespace nsDelegates
{
    using System;

    class clsDelegate
    {
        public delegate void TextHandler (string str);
        public TextHandler ShowText = null;

        public void ExecuteDelegate ()
        {
            if (ShowText != null)
                ShowText ("Delegate called");
        }
    }

    class clsMain
    {
        public static void Main ()
        {
            clsDelegate dlg = new clsDelegate();
            dlg.ShowText =
                new clsDelegate.TextHandler (Static);
            dlg.ExecuteDelegate();

            clsMain main = new clsMain ();
            dlg.ShowText =
```

```
            new clsDelegate.TextHandler (main.Instance);
        dlg.ExecuteDelegate();
    }

    private static void Static(string str)
    {
        Console.WriteLine ("Static() Method: " + str);
    }
    private void Instance(string str)
    {
        Console.WriteLine ("Instance() Method: " + str);
    }
    }
}
```

The *DlgClass.cs* program achieves the same result as *SinglCst.cs* in the last section and produces the same output, except now the delegate is type-safe. The *ExecuteDelegate()* function may not call any member methods in *clsMain* except the method you assign to *ShowText*. In addition, you may declare and use an instance of *clsDelegate* in any other class.

Using Delegates with Arguments and Return Values

The delegates in the previous examples have had empty parameter lists and do not return any values. This is not always the case. In fact, a more common use for delegates is to pass information to the delegate to perform an operation. Very often, the delegate will return the result of that operation.

The following program, *MathDlg.cs*, uses delegates to successively return the sum, difference and product of two numbers:

```
namespace nsMathDelegate
{
    using System;
    class clsDelegate
    {
        public delegate int MathHandler (int x, int y);
        public MathHandler DoMath;
```

```
        public int Invoke (int x, int y)
        {
            return (DoMath (x, y));
        }
    }

    class clsMain
    {
        public static void Main ()
        {
            clsDelegate dlg = new clsDelegate();
            clsMain main = new clsMain ();
            dlg.DoMath =
                new clsDelegate.MathHandler (main.Sum);
            Console.WriteLine (dlg.Invoke(14, 3));
            dlg.DoMath =
                new clsDelegate.MathHandler(main.Difference);
            Console.WriteLine (dlg.Invoke(14, 3));
            dlg.DoMath =
                new clsDelegate.MathHandler (main.Product);
            Console.WriteLine (dlg.Invoke(14, 3));
        }

        private int Sum (int x, int y)
        {
            Console.Write ("{0} + {1} = ", x, y);
            return (x + y);
        }
        private int Difference (int x, int y)
        {
            Console.Write ("{0} - {1} = ", x, y);
            return (x - y);
        }
        private int Product (int x, int y)
        {
            Console.Write ("{0} X {1} = ", x, y);
            return (x * y);
        }
    }
}
```

Delegates May Return Only One Value

It seems there always is a catch. Delegates come in two flavors, **single-cast** delegates and **multi-cast** delegates. In the examples in the previous sections, the methods that you assigned to the delegate had a return type of **void**. The compiler then cast them to multi-cast delegates, which permitted you to use the **+=** operator to add multiple methods to the delegate.

In the original C# compiler, attempting to assign multiple methods by using the **+=** operator in **MathDlg.cs** would result in your program throwing a **MulticastNotSupportedException**. In the latest version of C#, your program will not throw an exception, but you will get the proper result only on the first call to **Invoke()**.

If you change the assignments in **MathDlg.cs** to **+=**, you will see the following output:

```
14 + 3 = 17
14 + 3 = 14 - 3 = 11
14 + 3 = 14 - 3 = 14 X 3 = 42
```

The first line is correct because you are invoking only one delegate function. However, in the second call, you are invoking two delegate functions, but printing only one value. In the third call, you invoke three delegate functions but print only one value.

Thus, the C# compiler now lets you do something you should not do. You need to be aware of this quirk and recognize this incorrect output.

Defining a Delegate for an Event

You used event handlers in Lesson 28, "Using Windows Controls," to intercept and use the mouse clicks on the buttons, radio buttons and check boxes. All of these event handlers had a return type of *void*. Events always are multi-cast delegates, and thus cannot return a value.

Instead of declaring your own delegates, you used the *Delegate* class indirectly. You added your event handlers using the *System.EventHandler* class. Actually, *System.EventHandler* is a subclass of the *MulticastDelegate* class.

To use *System.EventHandler*, your event handler must be of the following form:

```
void HandlerName (object sender, System.EventArgs e)
```

The *sender* parameter identifies the control that generated the event, and *e* contains the arguments that you may use with the event:

```
void HandlerName (object sender, System.EventArgs e)
{
    if (object == MyControl)
    {
        // Statements
    }
}
```

You must refer to the documentation for the control event to determine what event arguments are available.

WHAT YOU MUST KNOW

In this lesson, you learned how to create and use delegates. You learned how to add multiple delegates in a multi-cast delegate, and to use single-cast delegates to return a value to your program. You also learned how to define delegates for an event generated by a Windows control. In Lesson 30, "Using the Common Dialogs in a C# Program," you will learn about the Windows *common dialog* library and how to use common dialogs in your program. Before you continue with Lesson 30, however, make sure you have learned the following key concepts:

- A delegate is the object-oriented programming equivalent of function pointers.

- A delegate defines a method but does not implement the method.

- To use a delegate, you first must assign a reference to a method to the delegate.

- You may assign multiple methods to a delegate if the delegate does not need to return a value to your program. This is a "multi-cast" delegate.

- If you need to have a delegate return a value to your program, then you must use a "single-cast" delegate.

*t*o simplify common programming operations, Windows provides a common controls library that programmers may use with any programming language. Common controls present the user with a consistent interface between programs. In addition to the common controls, Windows also provides a common dialogs library that programmers may use to provide common operations such as selecting fonts or colors, or to open and save files. In this lesson, you will learn how to display and use some of the common dialog boxes in a C# Windows application. By the time you finish this lesson, you will understand the following key concepts:

- Windows contains a common dialog library that any programming language may use to display common dialog boxes, such as an Open or Save As dialog box.
- By using the common dialogs, program users have a consistent interface to common operations.
- The common dialog library contains dialog boxes that let users select a system font, select a system color, open and save files, and so on.
- C# implements the Open and Save dialog boxes as separate hidden controls for your form.
- The common dialog box hidden controls contain properties your program code can set to customize the dialog boxes.

Understanding the Common Dialogs

Regardless of an application's processing, most programs perform similar operations. For example, most applications must open and close files. To help programmers perform these operations, Windows provides a set of common dialog boxes that programmers can easily integrate into their code. The dialog boxes present users with a common interface to standard operations.

By presenting a common interface, the time it takes a user to learn a new program is less than if each application provided its own custom interface. After a user learns how to save a file in one program, the user knows how to save a file in every application that uses the common dialog boxes.

To demonstrate how to use some of the common dialog boxes, create a new C# Windows application named "Dialogs." Use the steps you learned in Lesson 26, "Writing a Windows Program" to create the application. To start, you must enlarge the form as shown in Figure 30.1.

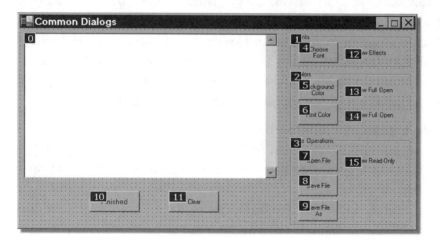

Figure 30.1. The Common Dialogs form that you will use to access Windows common dialogs.

Place the controls on the form as shown in Figure 30.1 to establish the "tab order." Add the large TextBox control to the form first. Draw the box to the size shown. Do not be concerned if the text box suddenly shrinks to a single line size. Show the Properties tool window and set the *Multiline* property to *true* to restore the text box to its full size.

Next, draw the three group boxes. Add the buttons and check boxes on the right, working your way down from the top. Finally, add the two buttons along the bottom from left to right.

The tab order is not important for this project, but if you make a mistake in adding the buttons and controls, you can change the tab order by setting the TabIndex property in the Properties tool window. Alternatively, you may select the View menu, then the Tab Order item and set the tab order by clicking on the controls. Neither way is easy, and definitely not nearly as easy as it was in Visual Studio 6. The TextBox should have a TabIndex of 0, assuring that it will have the "focus" (it is selected) when the form first displays.

If you have to change the tab order, notice that the Forms Designer assigns a sub-order to controls within a group box. For example, the Fonts group box has a tab order of 1, but the controls have tab orders of 1.0 and 1.1. This is new to Visual Studio and is a handy refinement. Use the Properties tool window to label the controls and name each control as shown in Table 30.1.

Control	Name Property	Text Property
textBox1	textBox1	<blank>
checkBox1	cbShowEffects	Show Effects
checkBox2	cbAllowFullOpen	Allow Full Open
checkBox3	cbShowFullOpen	Show Full Open
checkBox4	cbShowReadOnly	Show Read Only
button1	btnFont	Choose Font
button2	btnBkgndColor	Background Color
button3	btnTextColor	Text Color
button4	btnFileOpen	Open File
button5	btnFileSave	Save File
button6	btnFileSaveAs	Save File As
button7	btnFinished	Finished
button8	btnClear	Clear

Table 30.1. Set the name and text properties for the controls on the form as shown.

Double-click your mouse on each button on the form to add an event handler for the Click event. You will not need any event handlers for the text box. This will not be a full editing project, but it will show you how to use some of the common dialogs.

Using the Toolbar, add ColorDialog, FontDialog, OpenFileDialog and SaveFileDialog controls to the form. These are hidden controls. You will not need to change their names or properties. When you add these controls, the Visual Studio will display a window similar to Figure 30.2 below the form.

Set the Filter property for both the *openFileDialog1* and *saveFileDialog1* controls to the following to make the dialog boxes display only files with an extension of *.txt*:

```
"Text Files (*.txt)|*.txt"
```

Change the Text property for the form itself to "Common Dialogs." Build the program and give it a test-run to make sure the form appears as shown in Figure 30.1. Open the *Form1.cs* source-code file by right-clicking your mouse on the form and selecting View Text from the pop-up menu. At the top of the source file, add a *using* statement for the *System.IO* namespace. The block of *using* statements should appear as follows:

```
using System;
using System.Drawing;
using System.Collections;
using System.ComponentModel;
using System.Windows.Forms;
using System.Data;
using System.IO;
```

Add the following fields and *InitForm()* method to the code, and make the *Form1* class constructor read as follows:

```
public Form1()
{
    //
    // Required for Windows Form Designer support
    //
    InitializeComponent();
//
// Add the following statement to the constructor
```

```
//
            InitForm ();
      }
//
// strFileName holds the name of file currently open.
//
      private string strFileName = "";
//
// bReadOnly is to keep track of the open mode. Setting this
// to true will keep you from writing over a file.
//
      private bool bReadOnly = false;
//
// These are the custom colors that will show on the Color
// dialog when it first appears. Initially, they are all
// white
      int [] CustomColors = new int [16]
      {
          0x00ffffff, 0x00ffffff, 0x00ffffff, 0x00ffffff,
          0x00ffffff, 0x00ffffff, 0x00ffffff, 0x00ffffff,
          0x00ffffff, 0x00ffffff, 0x00ffffff, 0x00ffffff,
          0x00ffffff, 0x00ffffff, 0x00ffffff, 0x00ffffff
      };

      protected void InitForm ()
      {
          colorDialog1.CustomColors = CustomColors;
          textBox1.Focus ();
      }
```

This will be the basic form you will use to demonstrate some of the common dialogs. You will use the color dialog to set the text and background colors for the text box, and the font dialog to set the font and size of the text. Then, you will make it possible to open and display the contents of a file using C# streams. Finally, you will add code to save any changes to the text, and to save the text in a file using a different file name. The *CustomColors* field will save any custom colors you create in the color dialog box.

Using *ColorDialog*

The Color common dialog box gives you a stable, well-established way to choose a color. How you apply that color is up to your program. The dialog box simply presents a block of up to 64 colors and an optional "analog" field to create your own custom colors.

You may display the Color dialog box in two forms. One shows the 48 of the 64 basic colors that used to form the "palette" used by color displays before VGAs, AGPs and the like came around. The other 16 colors are "custom colors." The basic form is shown in Figure 30.3.

Figure 30.3. The basic Color dialog box.
The 16 boxes along the bottom are for custom colors
that you may create in your code or by selecting them
from the full color dialog box.

You can create the 16 custom colors in your code by declaring a 16-element int array. The lowest 24 bits of the integer make up the color, starting with red in the lowest 8 bits, green in the next and then blue:

```
0x0080D2F0
0x00 80 D2 F0
      /   |   \
   Blue  Green  Red
```

Setting all the colors to 0 will produce black. Setting all the colors to 255, or FF in hexadecimal, produces white. In this sample application, you set the Color dialog's *CustomColors* property to an array containing 16 colors, all white. The array is the *CustomColor* field in the *Form1* class. Try changing some of these values after you test-run the Color dialog box to see what colors different values produce.

The second form of the Color dialog box shows the advanced analog portion where you may create any of the colors your display is capable of showing. The full color dialog box is shown in Figure 30.4.

Figure 30.4. The full Color dialog box contains the basic portion on the left and the advanced portion on the right.

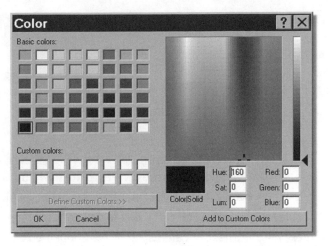

You may direct the Color dialog box to show the basic or full box when it first displays. In this project, you do that by setting the *ShowFullOpen* property of the dialog box object. Add the following code:

```csharp
protected void btnTextColor_Click (object sender,
                                   System.EventArgs e)
{
    colorDialog1.AllowFullOpen = cbAllowFullOpen.Checked;
    colorDialog1.FullOpen = cbShowFullOpen.Checked;
    colorDialog1.Color = textBox1.ForeColor;
    if (colorDialog1.ShowDialog () == DialogResult.Cancel)
    {
        return;
    }
    if (colorDialog1.Color == textBox1.ForeColor)
    {
        return;
    }
    textBox1.ForeColor = colorDialog1.Color;
}

protected void btnBkgndColor_Click (object sender,
                                    System.EventArgs e)
{
    colorDialog1.AllowFullOpen = cbAllowFullOpen.Checked;
    colorDialog1.FullOpen = cbShowFullOpen.Checked;
    if (colorDialog1.ShowDialog () == DialogResult.Cancel)
    {
        return;
    }
```

```
    if (colorDialog1.Color == textBox1.BackColor)
    {
        return;
    }
    textBox1.BackColor = colorDialog1.Color;
}
```

In each of these two event handlers, you use the check box values to set the *FullOpen* and *AllowFullOpen* properties. When the Show Full Open box is checked, the full dialog box appears and your program ignores the Allow Full Open check box. If, however, the Show Full Open box is not checked, the Allow Full Open box determines whether the user can display the full open box. The dialog will appear in its basic form, and allowing full open permits the user to use the Show Custom Colors button. Otherwise, the Show Custom Colors button is disabled.

When you compile and test-run the program, try typing some text in the text box. With some text, try changing the background and text colors to see what happens. Later, you will be able to display the contents of a file within the text box.

Common Dialogs are Information Devices

*When you use the common dialogs, you must remember that the dialog boxes themselves do **not** perform the operations they are designed to implement. Instead, they are information-gathering devices that you may customize then use to retrieve information from a user.*

For example, the Open dialog box does not actually open a file. It simply gives you a device from which the user can select a file name conveniently. The Color dialog does not actually set the color for anything. You use the Color dialog box so the user can select a color, which you then must apply to an object.

By using the common dialog boxes, you save yourself a lot of programming. Like the common controls, if they did not exist, you probably would want to write your own dialog box library to implement the common dialog. Microsoft has written the library for its programs and made the library available to you.

Using *FontDialog*

A typical personal computer has many fonts installed on it, and some may have several hundred. The *FontDialog* class presents a common interface for displaying these fonts and giving the user a consistent method of selecting fonts across applications. The *FontDialog* class displays the Font common dialog box. This is the same Font dialog box that Visual Studio displays when you select and modify the *Font* property of a form or control. The Font dialog is shown in Figure 30.5.

Figure 30.5. The Font common dialog box has options to set the effect and point size of a font.

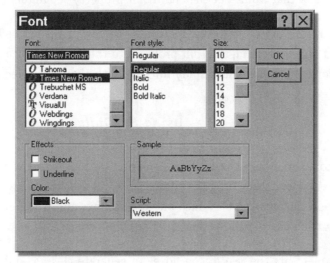

You can display the Effects section of the Font dialog box by setting the *ShowEffects* property to true. Within the effect section, you may display or hide the color box by setting the *ShowColor* property. To show the color box, however, you must set both the *ShowColor* and *ShowEffects* properties to *true*.

The code to implement the Font dialog and apply its result to the text box on the Form is shown in the following listing for the *btnFont* button Click event handler:

```
protected void btnFont_Click (object sender,
                              System.EventArgs e)
{
    fontDialog1.ShowEffects = cbShowEffects.Checked;
    fontDialog1.ShowColor = cbShowEffects.Checked;
    fontDialog1.Font = textBox1.Font;
    if (fontDialog1.ShowDialog () == DialogResult.Cancel)
    {
        return;
    }
    if (fontDialog1.Font == textBox1.Font)
```

```
    {
        return;
    }
    textBox1.Font = fontDialog1.Font;
}
```

Notice that the code sets both the *ShowEffects* and *ShowColor* properties according to the state of the check box next to the button. You could add a second check box that would set the *ShowColor* property separately.

In addition, notice that the code sets the *Font* property before displaying the dialog box. By setting the *Font* property to the same font that the text box uses, the Font dialog box will display initially with that font selected.

Using *FileDialog*

The common dialog box that implements the file open and file save operations comes in two forms, an Open dialog box and a Save As dialog box. Within the common dialogs, there is only one dialog box, and whether it is an open or a save dialog box depends upon the flags that you set.

C# implements the two dialog boxes in separate classes, and the Form Designer in Visual Studio contains separate hidden controls for each dialog box, an OpenFileDialog control and a SaveFileDialog control. The Open dialog box is shown in Figure 30.6.

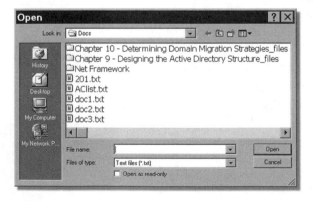

Figure 30.6. The Open dialog box is a common interface for opening files in Windows.

Notice that the dialog box in Figure 30.6 contains a box near the bottom center labeled "Open as read-only." This box is for the programmer's use. Windows does not actually open the file for you when the user selects a file and clicks on the Open button. Instead, the dialog box is simply an information-gathering device. After the user clicks his or her mouse on one of the buttons, you read the contents of the dialog box to get the selection information. If the user clicked on the Cancel button, you should ignore the information.

When the user closes the dialog box, your code is responsible for opening the file in the proper mode. In this project, the program simply opens the file, reads the contents and then closes the file. So the program maintains a *bReadOnly* field to hold the state of the read-only box. If you keep the file open, you may block other programs from using the file, but that is the approach many programs take. The code to implement the Open dialog box is shown in the following listing:

```csharp
protected void btnFileOpen_Click (object sender, System.EventArgs e)
{
// Initialize with the last file name used.
    openFileDialog1.FileName = strFileName;
// The default extension is for text files
    openFileDialog1.DefaultExt = ".txt";
    openFileDialog1.ShowReadOnly = cbShowReadOnly.Checked;
    if (openFileDialog1.ShowDialog () ==
                                    DialogResult.Cancel)
    {
        return;
    }
    this.strFileName = openFileDialog1.FileName;
    FileStream strm;
    try
    {
        strm = new FileStream (strFileName,
                                FileMode.Open,
                                FileAccess.Read);
        StreamReader reader = new StreamReader(strm);
//
// You do not need the loop, but Peek() tells you when you
// have reached the end of the file. To read the file line
// by line, you will need the loop.
        while (reader.Peek () > 0)
        {
            string str = reader.ReadToEnd();
            textBox1.Text = str;
```

```
      }
// Save the state of the read only box
      bReadOnly = openFileDialog1.ReadOnlyChecked;
      strm.Close ();
// Adding text to the text box sets its Modified property to
// true. The file has not actually been modified, so reset
// the Modified property.
      textBox1.Modified = false;
    }
// Catch the exception when the file cannot be found.
   catch (FileNotFoundException)
   {
      MessageBox.Show ("Cannot open file", "Warning");
   }
}
```

Compile and test-run your program. You should be able to open a file and display it in the text box. You still cannot save any changes. That will be the next step when you use the Save As common dialog box.

The Save As dialog box, shown in Figure 30.7, does not have the Read Only check box. If you are writing a file, then a read-only flag is meaningless.

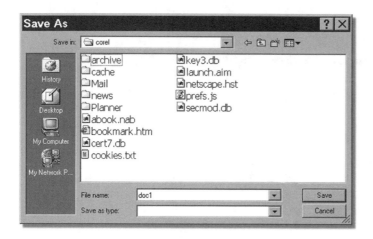

Figure 30.7. The Save As common dialog is the user's interface to saving a file in Windows.

You may initialize the File Name box by setting the *FileName* property to a string containing the file name. If this property is empty or blank, C# will pass a string containing "doc" and a sequence number to the dialog box. In this project, if you open a file and then press the Save File As button, the dialog will display with the original file name. You may change the name in the File Name box. The code to implement the Save As dialog box is shown in the following listing:

```csharp
protected void btnFileSaveAs_Click (object sender,
System.EventArgs e)
{
    if (strFileName.Length > 0)
    {
        saveFileDialog1.FileName = strFileName;
    }
    if (saveFileDialog1.ShowDialog () ==
                                    DialogResult.Cancel)
    {
        return;
    }
    if (saveFileDialog1.FileName.Length == 0)
    {
        return;
    }
    string fn = saveFileDialog1.FileName;
    FileStream strm;
    try
    {
        strm = new FileStream (fn, FileMode.OpenOrCreate,
                                FileAccess.Write);
        StreamWriter writer = new StreamWriter(strm);
        writer.Write (textBox1.Text);
        writer.Flush ();
        // Chop off any straggler text
        strm.SetLength (textBox1.Length);
        strm.Close ();
        textBox1.Modified = false;
        strFileName = fn;
    }
    catch (FileNotFoundException)
    {
        MessageBox.Show ("Cannot open file", "Warning");
    }
    catch (NotSupportedException)
    {
        MessageBox.Show ("Cannot write to file", "Warning");
    }
    catch (UnauthorizedAccessException)
    {
        MessageBox.Show ("Not authorized to write to file",
```

```
                                    "Warning");
    }
}
```

As you can see, this code is a bit longer than the code for the FileOpen button. That's because there are more possibilities for your program to throw an exception.

One last possibility you must consider is when the user opens a file, makes some changes and then wants to save the file. You do not want to go through the SaveFile dialog for every change. You already know the file name, so you can just write the contents. That is the purpose of the Save File button. The following function is the event handler for that button:

```
protected void btnFileSave_Click (object sender,
                                    System.EventArgs e)
{
    if (bReadOnly)
    {
        MessageBox.Show ("File is open as Read-Only",
                        "Warning");
        return;
    }
    if (strFileName.Length == 0)
    {
        btnFileSaveAs_Click (sender, e);
        return;
    }
    if (textBox1.Modified == false)
    {
    return;
    }
    if (strFileName.Length == 0)
    {
        return;
    }
    FileStream strm;
    try
    {
        strm = new FileStream (strFileName,
                                FileMode.OpenOrCreate,
                                FileAccess.Write);
```

```
            StreamWriter writer = new StreamWriter(strm);
            writer.Write (textBox1.Text);
            writer.Flush ();
            // Chop off any straggler text
            strm.SetLength (textBox1.Length);
            strm.Close ();
            textBox1.Modified = false;
    }
    catch (FileNotFoundException)
    {
        MessageBox.Show ("Cannot open file", "Warning");
    }
    catch (NotSupportedException)
    {
    MessageBox.Show ("Cannot write to file", "Warning");
    }
    catch (UnauthorizedAccessException)
    {
    MessageBox.Show ("Not authorized to write to file",
                    "Warning");
    }
}
```

If the file is open as read only, you warn the user and then exit. At this point, the user has the option of using the Save File As button to save the file under a new name, but you do not want to overwrite a file that you opened as read only.

In addition, you have several exception possibilities to catch in this function. You can continue "stacking" the exceptions by writing different *catch* blocks for the exceptions. You still have a couple of other buttons that you need to write code for: the Finished button and the Clear button. The code for these event handlers is shown in the following listing:

```
protected void btnClear_Click (object sender,
                               System.EventArgs e)
{
//
// If the text has changed since the last save, ask first
// before clearing the text
    if (textBox1.Modified == true)
```

```
    {
        string str;
        if (strFileName.Length > 0)
        {
            str = strFileName;
        }
        else
        {
            str = "The text in the edit control ";
        }
        str += " has changed. Do you want to save it?";
        if (MessageBox.Show(str, "TextChanged",
                    MessageBox.YesNo) == DialogResult.Yes)
        {
            btnFileSave_Click (sender, e);
        }
    }
//
// Erase any text in the text box.
    textBox1.Clear ();
// and set the modified flag to false
    textBox1.Modified = false;
}

protected void btnFinished_Click (object sender,
                                  System.EventArgs e)
{
    btnClear_Click (sender, e);
    Close ();
}
```

The Clear button event handler first checks whether the text has changed, and if so asks the user whether to save it first. This button is the equivalent of the New menu item in *NotePad*. When you press the Finished button to end the program, the event handler first calls the Clear button event handler. Again, this gives the user a chance to save any changes before the text is lost.

Using the Common Dialogs Without a Form

*Using a Windows form-based application makes it very easy to use the common dialogs. You simply select the dialog box control on the Toolbox tool window and drop it onto your form. Then, you simply call the control's **ShowDialog()** method to display the dialog box.*

*You do not need to use a forms-based application to use the common dialog boxes, however. You can implement the common dialog boxes from a command line program as well. The following program, **CmdLine.cs**, runs from the command line and displays the Color dialog box:*

```
namespace nsCommonDialog
{
    using System;
    using System.Drawing;
    using System.Windows.Forms;
    class clsMain
    {
        static public void Main ()
        {
            ColorDialog color = new ColorDialog();
            color.ShowDialog ();
        }
    }
}
```

*To compile **CmdLine.cs**, use the following command:*

```
C>csc cmdline.cs
```

WHAT YOU MUST KNOW

In this lesson, you learned how to add common dialog boxes to your program. You learned why you should use common dialog boxes to present a common interface when the user switches between programs. In addition, by using the common dialog boxes, you save yourself a lot of programming. You also learned how to set the properties and to display the common dialog boxes in your program, and how to retrieve information from the properties. In Lesson 31, "Using Streams in C#," you will learn how to open, read from, write to and close files in C#. Before you continue with Lesson 31, however, make sure you have learned the following key concepts:

- You add common dialog boxes to your program by selecting the item on the Toolbox tool window and dropping it on your form.

- The common dialog box controls are "hidden" controls. They do not appear on your form, but the Visual Studio makes them a part of your project.

- You customize the common dialog boxes by setting properties. For example, you prevent the user from displaying the full Color dialog box by setting the *AllowFullOpen* property to *false*.

- You can use the properties to set basic parameters for the common dialog boxes as well. For example, by setting the *FileName* property of the Open dialog box, you cause the dialog box to display a default file name.

- When the user closes a common dialog box, C# sets the properties. You may read the properties to determine options the user selected.

USING STREAMS IN C#

*W*riting programs often means that you must read a file to retrieve information that your program will need, or that you must write to a file to save information to use in your program later or for use by other programs. At this stage of its development, file handling is not one of the stronger points of C#. In fact, when you create a C# Windows application, the Visual Studio does not even add the namespace that you need to use files; you must add that manually. However, C# does implement the same standard input, output and error *streams* used by C and C++. In addition, the C# stream classes enable you to read and write files. In this lesson, you will learn to use the stream classes, and how to read and write files. By the time you finish this lesson, you will understand the following key concepts:

- A *stream* is a data path from one part of your computer to another. The end points may be a device such as the keyboard or a file.
- C# implements the standard in, standard out and standard error streams using the *FileStream* class.
- The *StreamReader* and *StreamWriter* classes read and write a file in text mode. You use these classes to read and write lines of text to and from a file.
- The *Stream* class is the base class for most stream operations.
- All classes derived from the *Stream* class and all objects you create using the *Stream* class inherit basic properties and methods that you may use to manipulate streams.

Understanding the Stream Classes

Very often your programs must read and write information to and from files. Until now, your programs have used the *Console* class to read from the keyboard and to write to the display. The *Console* class provides *stream* operations that make it easier to perform these functions. The *Console.Read()* accepts keyboard input from the user through a stream and the *Console.Write()* provides a stream to write to the display.

A *stream* is a sequence of data flowing from one part of your computer to another. For example, when you use the *Console.Read()* method in your program, data flows from the keyboard to your program. When you use the *Console.Write()* function, data flows from your program to the screen. Data written by a program travels downstream to its destination, and data read by a program travels upstream from its source.

C# implements the *Console* class using streams. You also may create streams that help you to read and write to and from files. When you do this, one end of the stream is your program and the other end is the file.

You can connect a stream from your program to a file by using the *FileStream* class. You can use the *FileStream* class directly. To read from or write to a file, you must use a *byte* array, as shown in the following program, *Writer.cs*, which reads lines you type from the keyboard and writes them to a file named *Writer.txt*. Notice that to use stream classes you must declare that you are using the *System.IO* namespace:

```
namespace nsFileStream
{
    using System;
    using System.IO;
    class clsMain
    {
        static public void Main(string [] args)
        {
            FileStream strm = new FileStream ("Writer.txt",
                                   FileMode.OpenOrCreate,
```

```
                                      FileAccess.Write);
            string str;
            while (true)
            {
                str = Console.ReadLine ();
                if (str.Length == 0)
                    break;
                byte [] text = new byte [str.Length + 2];
                int x;
                for (x = 0; x < str.Length; ++x)
                    text[x] = (byte) str[x];
                text[x++] = (byte) '\r';
                text[x] = (byte) '\n';
                strm.Write (text, 0, text.Length);
            }
            strm.Close ();
        }
    }
}
```

When you compile and run this *Writer.cs*, type a couple of lines such as the following. Pressing the **Enter** key on an empty line ends the program. When the program exits, show the file to make sure the program wrote the text correctly.

```
C:> writer  <Enter>
Now is the time to come to the aid of the Teletype. <Enter>
The quick red fox jumps over the lazy brown dog. <Enter>
<Enter>

C:> type writer.txt  <Enter>
Now is the time to come to the aid of the Teletype.
The quick red fox jumps over the lazy brown dog.

C:>
```

You may read a file in the same way, accepting the text into a *byte* array. The following program, *Reader.cs*, reads the file you just wrote using *Writer*, and prints it to your screen:

```
namespace nsFileStream
{
    using System;
    using System.IO;
    class clsMain
    {
        static public void Main()
        {
            FileStream strm;
            try
            {
                strm = new FileStream ("Writer.txt",
                                        FileMode.Open,
                                        FileAccess.Read);
            }
            catch (FileNotFoundException)
            {
                Console.WriteLine ("Could not open file");
                return;
            }
            byte [] text = new byte [256];
            while (true)
            {
                int len = strm.Read (text, 0, text.Length);
                if (len <= 0)
                    break;
                int x;
                string str = "";
                for (x = 0; x < len; ++x)
                    str += (char) text[x];
                Console.Write (str);
            }
            strm.Close();
        }
    }
}
```

To use *FileStream*, you must specify the file name, an open mode, and an access mode. The access mode should be one of the flags listed in Table 31.1. Notice that *FileAccess.ReadWrite* is the same as *FileAccess.Read* | *FileAccess.Write*.

Access Flag	Meaning
Read	Data can be read from the file and the file pointer can be moved.
ReadWrite	Data can be written to the file and the file pointer can be moved. Data can also be read from the file.
Write	Data can be written to the file and the file pointer can be moved.

Table 31.1. File access flags used with the C# stream classes.

The file open mode must be one of the modes listed in Table 31.2. In addition, when using Windows NT and Windows 2000, you must have permission to open the file in the specified mode.

Open Mode	Meaning
Append	Opens the file if it exists and seeks to the end of the file. If the file does not exist, a new file will be created. You must use this mode only with FileMode.Append or an ArgumentException will be thrown.
Create	Specifies that the operating system should create a new file. If the file already exists, it will be overwritten. You must also specify FileAccess.ReadWrite or FileAccess.Read.
CreateNew	The operating system should create a new file. If the file exists, an IOException will be thrown. Either the FileAccess.Write or FileAccess.ReadWrite flag is required.
Open	Specifies that the operating system should open an existing file.
OpenOrCreate	The operating system should open a file if it exists or create a new file if the file does not exist.
Truncate	The operating system should open an existing file. Once opened, the file should be truncated so that its size is zero bytes. You must specify FileAccess.Write or FileAccess.ReadWrite.

Table 31.2. File open mode flags for use with C# stream classes.

Notice in Table 31.2 that you must use certain access flags when you specify certain open modes. Obviously, read access only when you create a new file or truncate an existing file is not very useful.

Using Stream Objects

C# uses the *FileStream* class to implement the standard input, standard output and standard error streams normally associated with the keyboard and screen. A *FileStream* object reads and writes files in bytes. In the *Reader.cs* sample from the last topic, you read the file in blocks until there was no more to read. The program did not read individual lines.

In addition, to write information, you needed to convert the 16-bit Unicode data used by the C# *char* and *string* type to eight-bit bytes before writing to the file. Then, when you read the file, you converted the characters back to Unicode.

Sometimes you must use a file as a byte stream, but very often in C# you must read individual lines of a file, process the one line and then go back and read the next line. When writing to a file, you often must write individual lines.

To process individual lines in a file, you may attach a *StreamReader* object to a stream to read a file line by line, or a *StreamWriter* object to write to the file line by line. Both the *StreamReader* and *StreamWriter* recognize individual lines. Both automatically convert between Unicode and eight-bit characters so that you may use the *string* and *char* data types directly.

To attach a *StreamWriter* object to a stream, create the stream first, then create the *StreamWriter* object specifying the stream in the constructor:

```
FileStream strm = new FileStream ("Writer.txt",
                                  FileMode.OpenOrCreate,
                                  FileAccess.Write);
StreamWriter writer = new StreamWriter (strm);
```

The following version of *Writer.cs* creates a *StreamWriter* object and attaches to it. The code is much simpler, and you may directly use the *string* object that stores the lines from the keyboard:

```
namespace nsFileStream
{
    using System;
    using System.IO;
    class clsMain
    {
        static public void Main(string [] args)
        {
            FileStream strm = new FileStream ("Writer.txt",
                                    FileMode.OpenOrCreate,
                                    FileAccess.Write);
            StreamWriter writer = new StreamWriter (strm);
            string str;
```

```
            while (true)
            {
                str = Console.ReadLine ();
                if (str.Length == 0)
                    break;
                writer.Write (str + "\r\n");
            }
            strm.Close ();
        }
    }
}
```

You use the *StreamReader* class similarly. First, create the *FileStream* object, then create the *StreamReader* object specifying the stream in the constructor. The following version of *Reader.cs* uses a *StreamReader* to simplify the code:

```
namespace nsFileStream
{
    using System;
    using System.IO;
    class clsMain
    {
        static public void Main()
        {
            FileStream strm;
            try
            {
                strm = new FileStream ("Writer.txt",
                                       FileMode.Open,
                                       FileAccess.Read);
            }
            catch (FileNotFoundException)
            {
                Console.WriteLine ("Cannot open file");
                return;
            }
            StreamReader reader = new StreamReader (strm);
            while (reader.Peek() > 0)
            {
                string str = reader.ReadLine ();
                str += "\r\n";
                Console.Write (str);
```

```
        }
        strm.Close();
    }
  }
}
```

You should notice that *StreamReader* removes the newline sequence when it reads a line from the file, making the line ready for parsing to extract the information from the *string* object. To write the line to the screen, you must add the line-ending sequence "\r\n" to the string.

To demonstrate parsing a line from a file in your program, create a new C# Windows application named "States." This program will read lines from a text file. Each line contains the name of a state, a comma and the name of the state's capital city. You then will parse the string to place the state name and the capital city in separate strings, which you then will add to a ListView control.

After you create the project, add a ListView control as shown in Figure 31.1.

Figure 31.1. The States project form has only a ListView control and a Label control.

Next, add a Label control below the ListView control. Make the Label control large enough to hold a couple of lines of text. Select the ListView control in the Form Editor and prepare the ListView control by using the following steps to modify the properties in the Properties tool window.

❶ Set the MultiSelect property to false and the FullRowSelect property to true.

❷ Change the View property to Details.

❸ In the Behavior section of the Properties tool window, click on the Columns item. A small button with an ellipsis will appear on the right of the entry. Click on the button to open the ColumnHeader Collection Editor.

❹ On the ColumnHeader Collection Editor, click the Add button twice to add two columns to the control. Click on the first entry in the Members box, then change the Text property to "State" in the Properties box. Click on the second entry and change the Text property to "Capital." Set the Width property of both columns to 120.

❺ After you make the changes to the column headers, click the OK button to return to the form. You should see the text in the column headers in the Forms Designer.

❻ Select the Events listing by clicking on the lightning bolt icon at the top of the Properties tool window. Scroll to find the SelectedIndexChanged event and select it. Type **listView1_SelectedIndexChanged** and press the **Enter** key. This will add an event-handler function where you will read the current selection and display it in the Label control.

Select the Label control and change the Text property so that it is blank.

Right-click your mouse anywhere on the form and select View Text from the menu that pops up. The Visual Studio will open the *Form1.cs* source file in the text editor. At the top of the source file, add the following line to the list of *using* statements:

```
using System.IO;
```

Although the Visual Studio does not add this line, you should make it a practice to do this, even if you do not plan to access files in your program. If you do not access files, the line will not add any code to your program, but if you later add stream objects and forget to add the line, you will get several compiler errors.

Next add the following line to the *Form1* class constructor. The construction should look like the following:

```
public Form1()
{
    //
    // Required for Windows Form Designer support
    //
    InitializeComponent();
    InitForm ();
    //
    // TODO: Add any constructor code after
    // InitializeComponent call
    //
}
```

Finally, you must add the *InitForm()* function and add code to the event handler for the ListView control. The code should read as shown below:

```
void InitForm ()
{
    FileStream strm;
    try
    {

        strm = new FileStream ("..\\States.txt",
                               FileMode.Open,
                               FileAccess.Read);
        StreamReader reader = new StreamReader (strm);
        while (reader.Peek() > 0)
        {
            string State = reader.ReadLine();
            string[] Data = State.Split (new char [] {','});
            ListViewItem item = new ListViewItem (Data);
            listView1.Items.Add (item);
        }
        strm.Close ();
    }
    catch (FileNotFoundException)
    {
        MessageBox.Show ("Cannot open file", "Warning");
    }
}

private void listView1_SelectedIndexChanged(object sender,
                                         System.EventArgs e)
{
    if (listView1.SelectedItems.Count == 0)
    {
        return;
    }
    string State =
            listView1.SelectedItems[0].SubItems[0].Text;
    string Capital =
            listView1.SelectedItems[0].SubItems[1].Text;
    if (State.Length > 0)
    {
        label1.Text = "The capital of " + State + " is "
                        + Capital;
    }
}
```

The *States.txt* file containing the data you will need is in the project's *bin* directory when you extract the project from this book's Web site. Placing it here makes it accessible from both Debug and Release versions of the program without changing the code.

When you compile and run *States.cs*, you will see the form in Figure 31.2.

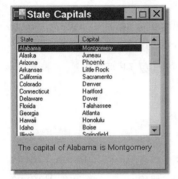

Figure 31.2. The States form displays the information from a file in a ListView control.

In the code, you first open the stream and then attach a *StreamReader* object to the stream. This makes it possible for you to read the file line by line, and also strips off the line ender from the line you read. The reading continues until the end of the file, when *reader.Peek()* returns a –1.

In the loop, you split the string into two parts using the *String* class *Split()* method. Comma-delimited data files are common, and this function will parse a string based on the character in the string you pass it.

You perform all this in a *try* statement. If your program cannot open the file, the code will display a message box.

Understanding Stream Properties

The *Stream* class is an *abstract* class and you may not create an object directly from *Stream*. You must create your stream objects from classes that are derived from *Stream*, such as the *FileStream* class. However, the basic *Stream* class does contain several properties to help you manipulate and use files. The properties are read-only and you cannot set them from your program.

All the stream classes inherit these properties. Table 31.3 summarizes the *Stream* properties.

Property	Use
CanRead	Indicates whether the current stream supports reading.
CanSeek	Indicates whether the current stream supports seeking.
CanWrite	Indicates whether the current stream supports writing.
Length	Returns the length in bytes of the stream.
Position	Indicates the position within the current stream.

Table 31.3. Properties that you may use when you create a stream object.

You may want to check the *CanRead* and *CanWrite* properties before you try to read and write a file. If, for example, the *CanWrite* property is *false* and you try to write to the file, your program will throw a *NotSupportedException*.

The *Length* property is useful when you want to create a string or byte array large enough to hold the entire file.

A stream object holds a value that indicates the position in the stream where the next character will be read or written. The value is in the *Position* property. You may change the *Position* property by using the *Seek()* method.

Using Stream Methods

Objects created from the *Stream* class, and other classes derived from *Stream*, inherit some basic methods for reading, writing, and moving around in a file. Derived classes may provide other methods, but the *Stream* methods always will be available.

After you successfully have opened a file in a stream, you may use any of the methods summarized in Table 31.4.

Method	Use
Close()	Closes the current stream and releases any resources (such as sockets and file handles) associated with the current stream.
Flush()	Updates the underlying data source or repository with the current state of the buffer and then clears the buffer. If the stream does not have a buffer, this method does nothing.
Read(buffer, offset, length)	Reads a sequence of bytes from the current stream and advances the current position within the stream by the number of bytes read.
ReadByte()	Reads a byte from the current position in the stream, or -1 if at the end of the stream.
Seek(offset, origin)	Sets the position within the current stream.
SetLength(new length)	Sets the length of the current stream.
Write(buffer, offset, count)	Writes a sequence of bytes to the current stream and advances the current position within this stream by the number of bytes written.
WriteByte(byte)	Writes a byte to the current position in the stream.

Table 31.4. Methods inherited by Stream objects and classes derived from Stream.

You should notice that the *Read()* and *Write()* methods require a *byte* array as a buffer, and also an offset and a length. The *offset* parameter is an offset from the beginning of the array, not an offset for the file position. You need to remember this when you read and write a file using the *Length* property of the array as the *count* parameter. For example, if you specify an offset of 2, the *Write()* method will begin adding data in the third byte of the array. If you use the array length as the *count* parameter, you will overwrite the array bounds and your program will throw an exception. The *Read()* method will begin placing data in the third byte of the buffer and will try to access two bytes beyond the array bound. Again, your program will throw an exception. The following line shows an incorrect statement:

```
byte [] data = new byte[24];
strm.Read (data, 2, data.Length);
```

To use this correctly, your program would have to specify *data.Length—2* as the count. The first two bytes of the array would retain their original values.

In addition, the *Read()* method appears to operate inconsistently. On the first read, it will begin placing data from the current file position into the third byte of the array. On successive reads, however, the method will advance the file position by the offset amount as well. Thus, you run the risk of losing data if you specify an offset.

Fortunately, for most file operations using the *Read()* or *Write()* methods, you will use an offset of 0.

The *Seek()* method requires a number of bytes to move within the stream. The *origin* parameter will determine where the seek will begin. You may use any of the following three values:

❶ SeekOrigin.Current. The seek operation will be from the current position in the stream. A negative value will move toward the beginning of the file and a positive value will move toward the end of the file.

❷ SeekOrigin.Begin. The seek operation will start at the beginning of the stream. If successful, the file pointer—the *Position* property—will equal the offset parameter for this method.

❸ SeekOrigin.End. The seek operation will begin at the end of the stream. If successful, the file pointer will equal the size of the file minus the offset parameter.

When using the *SetLength()* method, the stream size will be set to the value you specify, regardless of the original size of the file. If the file is larger than the value you specify, this method will truncate the file. If the file is smaller than the value you specify, *SetLength()* will pad the end of the file with zeros.

WHAT YOU MUST KNOW

In this lesson, you learned about streams in C#. To read from and write to a file, you first need to create an object using a class derived from the *Stream* class. The *Stream* class is an *abstract* class so you cannot create an instance of *Stream* directly. However, you learned that *Stream* contains the basic properties and methods that you will need to manipulate a stream. In Lesson 32, "Overloading Functions and Operators," you will learn how to write functions that have the same name, and how to redefine the symbols that C# uses for operators. Before you continue with Lesson 32, however, make sure you have learned the following key concepts:

✗ When you create a stream, data flows from one end of the stream to the other. The end points may be a device such as the keyboard or a modem or a file.

✗ The *FileStream* class is derived from the *Stream* abstract class. You use *FileStream* to create stream objects.

✗ By attaching a *StreamWriter* or *StreamReader* object to a stream, you may translate text files to and from Unicode to eight-bit characters.

✗ The *Stream* class holds a number of properties that you may read to determine the status of streams, and methods that you may use to manipulate streams.

✗ Other classes derived from *Stream* may provide other properties and methods, but the *Stream* properties and methods will be available in any derived class.

LESSON 32

OVERLOADING FUNCTIONS AND OPERATORS

*i*n C#, the functions in a class must be unique. This does not mean, however, that the names of the functions must be unique. The compiler prepares a *signature* of each function and compares them with the signatures of other functions. If any two signatures are identical, the compiler issues an error. Two functions may have the same name but have different signatures through a process known as *overloading*. In addition, you may use overloading to redefine C# operators for special purposes. In this lesson, you will learn how to overload functions and operators. By the time you finish this lesson, you will understand the following key concepts:

- The C# compiler creates a signature for each function in a class. The signature for the function must be unique within the class.
- The compiler creates the signature using the function name and the number and data types of the function's parameters. The return type of the function is not part of the signature.
- Two functions may have the same name, but have unique signatures.
- Operators as well as functions may be overloaded to provide special operations when used with C# objects.

Understanding Overloading

When you compile a program in C#, the compiler creates a *signature* for each function. The signature is a combination of the fully-qualified name of a function, including the namespace name and the class name, and its parameter list. If the compiler encounters two or more functions with identical signatures, it cannot tell the difference between them and it cannot tell which function your program intends to use. The compiler will then issue an error message and your program's compilation will fail.

The signatures of two functions may be different even though the fully qualified name is the same. The Visual C# compiler will let you declare and define both functions. This is because it can identify the functions as unique because of their signatures.

This process is *function overloading*, and it lets you define and use functions using the same name but which have different parameter lists. For two functions with the same name, the parameter list must differ in the number or data type of the parameters. The following three functions use the same name, *Func*, but their signatures are different because the parameter differs in the number or data type of the parameters:

```
void Func (int x)
{
}
void Func (int x, int y)
{
}
void Func (double x)
{
}
```

Neither the return type nor the names of the parameters are part of a function's signature. The return type is not a part of the signature because Visual C# lets you discard a return value by not assigning it to a variable, and thus the compiler cannot know which function you intend to call. The parameter names are not a part of the signature because Visual C# lets you call functions with variables that have names different from the parameter names.

Normally, you would use overloaded functions to perform similar operations on different data types. However, C# does not require that your functions perform similar operations. Overloaded functions could perform very different operations, but that would only tend to make your programming confusing.

The following program, *Overload.cs*, shows how you could use two functions, both named *Square()* to return the squares of an *int* data type and a *double* data type:

```csharp
namespace nsOverload
{
    using System;
    class clsMain
    {
        static public void Main ()
        {
            Console.WriteLine ("The square of 9 is "
                                + Square(9));
            Console.WriteLine ("The square of 5.2 is " +
                                "{0,0:F2}", Square(5.2));
        }

        static int Square (int val)
        {
            Console.WriteLine ("int Square(int val) " +
                                "called.");
            return (val * val);
        }

        static double Square (double val)
        {
            Console.WriteLine ("double Square(double val) "+
                                "called.");
            return (val * val);
        }
    }
}
```

When you compile and run this command line program, you should see the following output:

```
int Square(int val) called.
The square of 9 is 81
double Square(double val) called.
The square of 5.2 is 27.04
```

Even though the functions you called have the same name, *Square*, the compiler was able to distinguish between them by the data types of the parameter list.

Overloading Functions

To overload a function, simply declare and define a function with the same name as another function, but with a different parameter list. The return type of both functions may be the same, but the parameter list must have different data types or a different number of parameters.

A major difference between functions in C# and C++ is that you may not assign default values to parameters in C# as you can do in C++. In C++, you assign the default values when you declare the function prototypes. C#, however, does not use function prototypes and makes no distinction between the function declaration and the function definition.

This makes function overloading easier. In C++, it sometimes can be difficult to spot "overloaded" functions that the compiler sees as identical, as in the following:

```
void Func (int x, double y = 24.8);
void Func (int x, char y = 'c');
void Func (int x, int y = 2);
```

To the Visual C++, compiler, all three of these functions are identical and the compiler would issue an error. In C#, you must declare these functions as follows:

```
void Func (int x, double y)
{
}
void Func (int x, char y);
{
}
void Func (int x, int y);
{
}
```

Here, there is no question that the functions are overloaded.

Overloading Constructors

*You may overload any function in a class except the destructor, or **Finalize()**, function. The destructor function cannot have parameters, so it cannot be overloaded. However, you can overload constructors to provide parameters. This will extend the ways in which you may create class objects. For example, you may write a class definition such as the following:*

```
class clsEmployee
{
    public clsEmployee ()
    {
        m_Name = "";
    }
    public clsEmployee (string strName)
    {
        Name = strName;
    }
    }
    public string Name;
}
```

*This will let you create a class object with or without providing a value for the **Name** field, as shown below:*

```
clsEmployee employee = new clsEmployee();
clsEmployee employee = new clsEmployee("Employee Name");
```

You will find numerous instances of overloaded constructors in the C# library code.

Overloading Operators

Sometimes the C# operators do not perform exactly as you would like when dealing with classes and other reference-type objects. In this case, you may redefine many of the operators to perform the tasks you must do. This process is *operator overloading*.

The syntax for overloading an operator is to write the *public* access keyword followed by the return type of the function, the keyword *operator* followed by the operator symbol (the "+" in the following sample) and the parameter list:

```
public clsClassName operator +(clsClassName arg1,
                               clsClassName arg2)
```

The access should always be *public* because more than likely you must use this operator outside the class definition. In addition, the number of arguments should be the same as the number of operands the non-overloaded operator requires.

Operator overloading in C# is much more restrictive than it is in C++. You may overload almost any operator in C++, but in C# you may overload only the unary, binary, and comparison operators. You cannot overload any of the 11 assignment operators such a =, +=, /=, and so forth.

C# always creates class objects using reference-type variables. This coupled with the restrictions on which operators you may overload can lead to some problems. In the following *Employ1.cs* program, for example, you create two employee records (class objects), one initialized with a name. Assume that you have several fields and properties in the class that you want to copy (for the sake of space, this program shows only the employee name). Rather than copy every field and property in separate assignment statements, you write just one assignment, *emp2 = emp1*. This is a legal assignment in C#, and copies all the variables in *emp1* to *emp2*. Then you change the name in *emp2*, fully expecting each class object to contain a different employee name.

```csharp
namespace nsEmployee
{
    using System;
    class clsMain
    {
        static public void Main ()
        {
            clsEmployee emp1 = new clsEmployee
                                    ("Thomas Jefferson");
            clsEmployee emp2 = new clsEmployee();
            emp2 = emp1;
            emp2.m_Name = "George Washington";
            Console.WriteLine ("emp1 name = " +
                                emp1.m_Name);
            Console.WriteLine ("emp2 name = " +
                                emp2.m_Name);

        }
    }

    class clsEmployee
    {
        public clsEmployee ()
        {
            m_Name = "";
        }
        public clsEmployee (string strName)
        {
            Name = strName;
        }
        public string m_Name
        {
```

```
        get
        {
            return (Name);
        }
        set
        {
            Name = value;
        }
    }
    private string Name;
    }
}
```

When you compile and run *Employ1.cs*, you see the following output:

```
emp1 name = George Washington
emp2 name = George Washington
```

This output is quite likely not exactly what you had in mind. You assigned a new value only to *emp2.m_Name* and you would expect *emp1.m_Name* to contain "Thomas Jefferson."

The problem is that strings in C# are reference objects. When you assign one class variable to another, your program dutifully copies the addresses—the *references*—to the strings from one class to the other. When you give *emp2* a new name, you also are writing over the name in *emp1*.

The same thing happens in C++, but you can simply overload the assignment operator to create a new string for the second class object. However, C# does not let you overload any of the assignment operators. You could overload one of the binary operators such as the addition operator, but writing something like *emp2 + emp1* would be confusing. In addition, you would have to write statements to copy all of the fields and properties from one class object to the other.

However, that is exactly what you want to do, and then you can take advantage of a quirk in the way a C# program evaluates assignment operators. When you write something like *x += 2*, your program first evaluates the assignment part of the operator and sets a temporary location equal to 2. Then it evaluates the addition part, adding the value of *x* to the temporary value and placing the result in *x*. Even though you must write them as a single operator, a C# program evaluates the two symbols separately.

This is to your advantage. If you overload the addition operator, which you probably do not need for class object operations anyway, the program first will perform the assignment, dutifully copying all the fields and properties from one object to the other. Then it will evaluate your overloaded + operator, in which you will create a new string to hold the name. Add the following code to the *clsEmployee* class in *Employ1.cs* (the complete code is in *Employ2.cs*):

```
public static clsEmployee operator +(clsEmployee emp1,
                                     clsEmployee emp2)
{
    emp1.Name = String.Copy(emp1.Name);
    return (emp1);
}
```

Although you do not use the *emp2* parameter, the + operator is a binary operator and requires two operands. Thus, overloading it means that you must provide the second, but unused, parameter. Now return to the *Main()* function and change the assignment statement to read as follows:

```
emp2 += emp1;
```

Now when you compile and run the program, you should see the following output:

```
emp1 name = Thomas Jefferson
emp2 name = George Washington
```

This is what you expected from the original program. By letting the assignment operator perform its work first, you can write whatever changes you need to make in the + operator overload. It is less confusing then just using something

like *emp2 + emp1*. Be aware that in the preceding overload, the addition operator only makes a new object for *m_Name*. If you use the operator without the assignment operator, other variables in the class object will not be copied.

The simple addition of a few lines of code has redefined the operator, and made your programming a lot easier.

You should remember that overloading an operator overloads it for operations involving only objects created from the class definition. Other statements using the operator, including statements involving objects created from other classes, will not be affected.

WHAT YOU MUST KNOW

In this lesson, you learned about function and operator overloading. You learned how to create functions that use the same name but that are unique to the compiler. You also learned how to overload, or redefine, C# operators to make the operators perform the task you need. This is the last lesson in this book. You have learned the basics of programming in C#, and you should consider this just the beginning. You will learn to program by programming. Every mistake will teach you something, so you should not worry about making errors. At this point, make sure you have learned the following key concepts:

- Overloading a function involves writing a function by the same name but which differs in the number or data type of the parameters.

- You may overload any function in a class except the destructor, or *Finalize()* function.

- Overloaded constructors extend the ways in which you may create a class object.

- You may overload operators as well as functions. When you overload a C# operator, you are redefining the operator only for the class that contains the operator overload.

- C# lets you overload only the unary, binary and comparison operators. You may not overload the assignment operators.